REBUILDING IRAQ:
RESTORING IRAQ'S OIL
AND ELECTRICITY SECTORS

REBUILDING IRAQ: RESTORING IRAQ'S OIL AND ELECTRICITY SECTORS

G.A.O.

Novinka Books
New York

Copyright © 2008 by Nova Science Publishers, Inc.

All rights reserved. No part of this book may be reproduced, stored in a retrieval system or transmitted in any form or by any means: electronic, electrostatic, magnetic, tape, mechanical photocopying, recording or otherwise without the written permission of the Publisher.

For permission to use material from this book please contact us:
Telephone 631-231-7269; Fax 631-231-8175
Web Site: http://www.novapublishers.com

NOTICE TO THE READER

The Publisher has taken reasonable care in the preparation of this book, but makes no expressed or implied warranty of any kind and assumes no responsibility for any errors or omissions. No liability is assumed for incidental or consequential damages in connection with or arising out of information contained in this book. The Publisher shall not be liable for any special, consequential, or exemplary damages resulting, in whole or in part, from the readers' use of, or reliance upon, this material.

This publication is designed to provide accurate and authoritative information with regard to the subject matter covered herein. It is sold with the clear understanding that the Publisher is not engaged in rendering legal or any other professional services. If legal or any other expert assistance is required, the services of a competent person should be sought. FROM A DECLARATION OF PARTICIPANTS JOINTLY ADOPTED BY A COMMITTEE OF THE AMERICAN BAR ASSOCIATION AND A COMMITTEE OF PUBLISHERS.

LIBRARY OF CONGRESS CATALOGING-IN-PUBLICATION DATA

United States. General Accounting Office.
  Rebuilding Iraq : restoring Iraq's oil and electricity sectors / GAO.
      p. cm.
  Includes bibliographical references and index.
  ISBN 978-1-60456-434-1 (softcover : alk. paper)
  1. Petroleum industry and trade--Iraq. 2. Electric utilities--Iraq. 3. Postwar reconstruction--Iraq. 4. Iraq--Economic conditions--21st century. I. Title.
  HD9576.I72U55   2008
  333.793'209567--dc22
                                                           2008012140

Published by Nova Science Publishers, Inc. ✤ *New York*

CONTENTS

Preface	vii
Integrated Strategic Plan Needed to Help Restore Iraq's Oil and Electricity Sectors	1
Results in Brief	3
Background	7
The United States, Iraq, and Donors Have Funded Reconstruction of Iraq's Oil and Electricity Sectors, but Iraq's Future Needs Are Significant and Sources of Funding Uncertain	13
Iraq's Oil and Electricity Production Goals Have Not Been Met, Oil Production Figures May Be Overstated, and Iraq Faces Difficulties Sustaining Infrastructure	23
Major Challenges Hinder Efforts to Meet Iraq's Oil and Electricity Needs	35
Conclusion	45
Recommendations for Executive Action	47
Appendix I. Objectives, Scope, and Methodology	51
References	55
Index	61

PREFACE

While billions have been provided to rebuild Iraq's oil and electricity sectors, Iraq's future needs are significant and sources of funding uncertain. For fiscal years 2003 through 2006, the United States made available about $7.4 billion and spent about $5.1 billion to rebuild the oil and electricity sectors. The United States spent an additional $3.8 billion in Iraqi funds on the two sectors, primarily on oil and electricity sector contracts administered by U.S. agencies. However, according to various estimates and officials, Iraq will need billions of additional dollars to rebuild, maintain, and secure Iraq's oil and electricity sectors. The Ministry of Electricity estimates that about $27 billion will be needed to meet the sector's future rebuilding requirements; a comparable estimate has not been developed by the Ministry of Oil. Since the majority (about 70 percent) of U.S. funds has been spent, the Iraqi government and the international donor community represent important sources of potential funding. However, prospects of such funding are uncertain. First, the Oil and Electricity Ministries have encountered difficulties spending capital improvement budgets because of weaknesses in budgeting, procurement, and financial management. As of November 2006, the Ministry of Oil had spent less than 3 percent of its $3.5 billion 2006 capital budget to improve Iraq's oil facilities. Second, Iraq has not made full use of potential international contributions and it is unclear what additional financial commitments, if any, will be provided to Iraq's oil and electricity sectors as part of a new international compact (agreement), according to U.S. officials. As of March 2007, donors had committed $580 million in grants for the electricity sector and had offered loans for oil and electricity projects; however, Iraq has not accessed these loans in part due to concerns about its high debt burden.

WHY GAO DID THIS STUDY

Since 2003, the United States has provided several billion dollars in reconstruction funds to help rebuild Iraq oil and electricity sectors, which are crucial to rebuilding Iraq's economy. For example, oil export revenues account for over half of Iraq's gross domestic product and over 90 percent of government revenues. The U.S. rebuilding program was predicated on three key assumptions: a permissive security environment, the ability to restore Iraq's essential services to prewar levels, and funding from Iraq and international donors.

This report addresses (1) the funding made available to rebuild Iraq's oil and electricity sectors, (2) the U.S. goals for these sectors and progress in achieving these goals, and (3) the key challenges the U.S. government faces in these efforts.

WHAT GAO RECOMMENDS

This report recommends that the Secretary of State, in conjunction with relevant U.S. agencies and international donors, work with Iraqi ministries to develop an integrated energy strategy. State commented that the Iraqi government, not the U.S. government, is responsible for taking action on GAO's recommendations. GAO believes that these recommendations are still valid given the billions made available for Iraq's energy sector and the U.S. government's influence in overseeing Iraq's rebuilding efforts.

WHAT GAO FOUND

Billions have been provided to rebuild Iraq's oil and electricity sectors, but Iraq's future needs are significant and sources of funding uncertain. From fiscal years 2003 through 2006, the United States spent about $5.1 billion to rebuild the oil and electricity sectors. The United States also spent an additional $3.8 billion in Iraqi funds on these sectors. However, Iraq will need billions of additional dollars to rebuild these sectors. The Iraqi government and donors represent important sources of potential funding. However, the oil and electricity ministries have encountered difficulties spending capital improvement budgets because of weaknesses in budgeting and procurement practices and major security challenges. Moreover, Iraq has not made full use of potential international contributions. It is also unclear what additional financial commitments, if any, will be provided to Iraq's oil and electricity sectors as part of a new international compact.

Despite 4 years of effort and the substantial resources expended, production in both sectors has consistently fallen below U.S. program goals. In addition, State's estimate of Iraq's oil production levels may be overstated due to inadequate metering that does not allow precise measurement of crude oil production. The Iraqi government projects that it will be able to meet the demand for electricity in 2009. However, these projections assume that the Ministry of Electricity will be assured of the stable supply of the fuel needed for electricity generation, which has been lacking due to poor coordination between the oil and electricity ministries.

A variety of security, corruption, legal, and planning challenges have impeded U.S. and Iraqi efforts to restore Iraq's oil and electricity sectors. The challenging security environment and insufficient protection efforts have continued to place workers and infrastructure at risk. Corruption, smuggling, and other illicit activities result in revenue losses and low cost recovery. Furthermore, the Iraqi government has difficulty attracting foreign investment because, according to the World Bank, it lacks an adequate legal framework, including comprehensive hydrocarbon legislation that would govern distribution of future oil revenues and granting of exploration rights. Finally, although the oil and electricity sectors are mutually dependent, the Iraqi government lacks integrated planning for these sectors, which has led to inefficient management of the country's resources.

Iraqi Reported Crude Oil Production, Exports, and U.S. Goals, June 2003 through December 2006

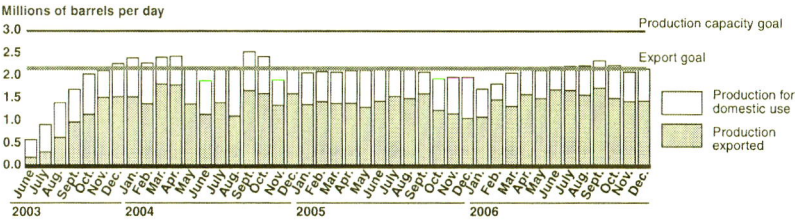

Source: Iraq Ministry of Oil estimates collected by State Department's Iraq Reconstruction and Management Office.

ABBREVIATIONS

CERP	Commander's Emergency Response Program
CPA	Coalition Provisional Authority
DAD	Donor Assistance Database
DFI	Development Fund for Iraq
DOD	Department of Defense
EIA	Energy Information Administration
EPSS	Electrical Power Security Service
GRD	Gulf Regional Division
IDA	International Development Assistance
IMF	International Monetary Fund
IRFFI	International Reconstruction Fund Facility for Iraq
IRMO	Iraq Reconstruction Management Office
IRRF	Iraq Relief and Reconstruction Fund
LPG	liquefied petroleum gas
mbpd	million barrels per day
mscfd	million standard cubic feet per day
mw	megawatt
mwh	megawatt hours
PCO	Project and Contracting Office
SIB	Strategic Infrastructure Battalions
SIGIR	Special Inspector General for Iraq Reconstruction
tpd	tons per day
UN	United Nations
UNDP	United Nations Development Programme
USAID	United States Agency for International Development

INTEGRATED STRATEGIC PLAN NEEDED TO HELP RESTORE IRAQ'S OIL AND ELECTRICITY SECTORS

May 15, 2007
Congressional Committees

The Iraqi government inherited an oil and electricity infrastructure that was greatly deteriorated due to the previous regime's neglect, international sanctions, and years of conflict, looting, and vandalism. Since 2003, the U.S. government has provided several billion dollars in reconstruction funds to help restore Iraq's crude oil production and exports and improve Iraq's electrical generating capacity, transmission, distribution, and monitoring systems.[1] Both sectors are crucial to rebuilding the economy. Iraq's crude oil reserves, estimated at 115 billion barrels, are the third largest in the world. Oil export revenues account for over half of Iraq's gross domestic product and over 90 percent of its revenues. This revenue is essential to Iraq's ability to provide for its needs, including reconstruction. In addition, an inadequate and unreliable supply of electricity affects both public perceptions of the government's ability to deliver basic services and the productivity of Iraq's oil sector, which is highly dependent on electricity.

GAO's review of reconstruction efforts in Iraq, done under the Comptroller General's authority to conduct evaluations on his own initiative, examined U.S. activities directed at rebuilding the oil and electricity sectors.[2] Specifically, we addressed (1) the funding made available to rebuild Iraq's oil and electricity sectors and the factors that may affect Iraq's ability to meet its future funding needs, (2) the U.S. goals for the oil and electricity sectors and progress in achieving these goals, and (3) the key challenges the U.S. government faces in helping Iraq restore its oil and electricity sectors.

To accomplish our objectives, we reviewed and analyzed U.S., Iraqi, donor government, United Nations (UN), International Monetary Fund (IMF), and World Bank reports and data. During our field work in Washington, D.C.; Baghdad, Iraq; and Amman, Jordan, we met with officials from the Department of State, the U.S. Agency for International Development (USAID), the Department of Defense (DOD), the Department of the Treasury, and the Department of Energy. During our two trips to Iraq and Jordan, we met with Iraqi, UN, IMF, World Bank, donor country (Japan and European Union), private sector, and U.S. officials. We observed several embassy meetings and a donor meeting led by the Iraqi Ministry of Planning. We also interviewed U.S., Iraqi, and UN officials at a November 2006 electricity conference sponsored by the UN Development Programme (UNDP) at the Dead Sea, Jordan. We analyzed data on Iraq's 2006 and 2007 budgets and 2006 budget expenditures. We found these data to be sufficiently reliable for comparing Iraq's intended ministry expenditures and determining that Iraq has not been able to fully expend its budget in certain budget categories. We examined data on Iraqi oil production from the Department of State's Iraq Reconstruction and Management Office (IRMO) and the Department of Energy's Energy Information Administration (EIA). Although we found some limitations in these data, they were sufficiently reliable to determine whether U.S. goals for oil production were being met. We also examined data on oil exports and electricity production in Iraq and found them sufficiently reliable for determining if U.S. goals were being met. Appendix I contains a more detailed description of our scope and methodology.

We conducted our review from January 2006 through March 2007 in accordance with generally accepted government auditing standards. The work performed for this review has also contributed to several related GAO products on Iraq.[3]

RESULTS IN BRIEF

While billions have been provided to rebuild Iraq's oil and electricity sectors, Iraq's future needs are significant and sources of funding uncertain. For fiscal years 2003 through 2006, the United States made available about $7.4 billion and spent about $5.1 billion to rebuild the oil and electricity sectors. The United States spent an additional $3.8 billion in Iraqi funds on the two sectors, primarily on oil and electricity sector contracts administered by U.S. agencies. However, according to various estimates and officials, Iraq will need billions of additional dollars to rebuild, maintain, and secure Iraq's oil and electricity sectors. The Ministry of Electricity estimates that about $27 billion will be needed to meet the sector's future rebuilding requirements; a comparable estimate has not been developed by the Ministry of Oil. Since the majority (about 70 percent) of U.S. funds has been spent, the Iraqi government and the international donor community represent important sources of potential funding. However, prospects of such funding are uncertain. First, the Oil and Electricity Ministries have encountered difficulties spending capital improvement budgets because of weaknesses in budgeting, procurement, and financial management. As of November 2006, the Ministry of Oil had spent less than 3 percent of its $3.5 billion 2006 capital budget to improve Iraq's oil facilities. Second, Iraq has not made full use of potential international contributions and it is unclear what additional financial commitments, if any, will be provided to Iraq's oil and electricity sectors as part of a new international compact (agreement), according to U.S. officials. As of March 2007, donors had committed $580 million in grants for the electricity sector and had offered loans for oil and electricity projects; however, Iraq has not accessed these loans in part due to concerns about its high debt burden.

Despite 4 years of effort and the substantial resources devoted to restoring Iraq's oil and electricity sectors, production in both sectors has consistently fallen below U.S. program goals. Key U.S. goals for the oil sector are to reach a crude oil production capacity of 3 million barrels per day (mbpd) and crude oil export levels of 2.2 mbpd, and to increase production capacity and stock levels of refined fuels. However, State Department data indicate that actual crude oil production and exports averaged, respectively, about 2.1 mbpd and 1.5 mbpd in 2006, and refined fuel production and stock levels did not meet the goals set. In addition, IRMO's estimate of Iraq's oil production levels may be overstated since inadequate metering does not allow precise measurement of crude oil production. The U.S. goal for electrical peak generation capacity is 6,000 megawatts (mw); however, electricity in Iraq averaged 4,280 mw of peak generation per day in 2006, about 3,950 mw short of demand in 2006. The Iraqi government projects that it will not be able to fully meet the demand for electricity until 2009. However, these projections assume that the Ministry of Electricity will be assured of a stable supply of the fuel needed for electricity generation, which has been lacking in the past due to poor coordination between the Oil and Electricity Ministries. Some improvements in current crude oil and electricity production levels may be achieved once the Army Corps of Engineers Gulf Region Division (GRD) finalizes its remaining oil and electricity projects, which are scheduled for completion by April 2008. However, these projects have experienced continued delays, undermining prospects for meeting U.S. production goals. Additionally, Iraq has experienced difficulty in sustaining the infrastructure rehabilitated with U.S. funds. For example, the U.S. decision to install gas turbine generators without an assured supply of natural gas required the use of fuel oils that resulted in increased maintenance costs and decreased power output.

A variety of security, corruption, legal, and planning challenges have impeded U.S. and Iraqi efforts to restore Iraq's oil and electricity sectors. These challenges have made it difficult to achieve the current crude oil production and export goals that are central to Iraq's government revenues and economic development. In the electricity sector, these challenges have made it difficult to achieve a reliable Iraqi electrical grid that provides power to all other infrastructure sectors and promotes economic activity. The U.S. reconstruction effort was predicated on the assumption that a permissive security environment would exist. However, the deteriorating security environment continues to place workers and infrastructure at risk while protection efforts have been insufficient. Corruption, smuggling, and other illicit activities result in revenue losses and low cost recovery. Widespread corruption and smuggling reduce revenues that could be collected through the legal sale of refined oil products such as gasoline. According to State Department officials and reports, about 10 percent to 30 percent of refined fuels is diverted to the black market or is smuggled out of Iraq and sold for a profit. Moreover, the lack of an effective metering system within the electricity sector hinders efforts to accurately measure consumption and deter theft. Furthermore, according to U.S. and World Bank officials, the government also lacks an adequate legal and regulatory framework, including comprehensive hydrocarbon legislation that would govern distribution of future oil revenues and granting of exploration rights. Until this framework is established, it will be difficult to attract the billions of dollars in additional foreign investment needed to modernize Iraq's oil sector and ensure that the government has the oil revenues essential for future reconstruction and reconciliation. Finally, although the oil and electricity sectors are mutually dependent, the Iraqi government lacks integrated planning for these sectors, which has led to inefficient management of the country's resources. According to the U.S. government, about $4 billion in potential revenues are lost each year through inefficient energy production practices, such as the flaring of natural gas, underscoring the need for such a plan.

This report makes several recommendations, including that the Secretary of State, in conjunction with relevant U.S. agencies and in coordination with the donor community, work with the Ministries of Oil and Electricity to (1) develop an integrated energy strategy for the oil and electricity sectors that identifies, among other items, key goals and priorities, future funding needs, and steps for enhancing ministerial coordination; (2) set milestones and assign resources to expedite efforts to establish an effective metering system for both the oil and electricity sectors; (3) improve the existing legal and regulatory framework, for example, by developing fair and equitable hydrocarbon legislation, regulations, and implementing guidelines; (4) set milestones and assign resources to expedite efforts to develop adequate budgeting, procurement, and financial management systems; and (5) implement a viable donor mechanism to secure funding for Iraq's future oil and electricity rebuilding needs.

In commenting on a draft of this report, State agreed that all the steps we included in our recommendations are necessary to improve Iraq's energy sector but stated that these actions are the direct responsibility of the Government of Iraq, not of the Department of State, any U.S. agency, or the international donor community. State also commented that U.S. agencies are already taking several actions consistent with our recommendations. We recognize that these actions are ultimately the responsibility of the Iraqi government. However, it is clear that the U.S. government wields considerable influence in overseeing Iraq stabilization and rebuilding efforts. We also acknowledge that State has made some efforts to address the issues that led to our recommendations and have refined our report accordingly. However, we believe additional actions are warranted given the lack of progress that has been made over the last 4 years in achieving Iraq reconstruction goals. Also, some of State's initiatives are relatively new. For example, State commented that an Energy Fusion Cell composed of the Embassy, Multinational Force-Iraq, and the ministries of oil and electricity has been formed to craft an integrated energy strategy and that the practical effectiveness of this effort will depend on whether the cell is able to obtain buy-in from the two Iraqi ministries. However, since the details of the strategy have not been released, it is still unclear whether this strategy will fully address the key elements of an effective strategy identified in our report, such as identifying rebuilding priorities, resource needs, stakeholder roles and responsibilities, and performance measures and milestones. The uncertainty over ministry buy-in also underscores the importance of State working closely with other U.S. agencies, the government of Iraq, and the international donor community in crafting a new energy sector strategic plan.

BACKGROUND

Iraq's oil infrastructure is an integrated network that includes crude oil fields and wells, pipelines, pump stations, refineries, gas/oil separation plants, gas processing plants, export terminals, and ports (see fig. 1). This infrastructure has deteriorated significantly over several decades due to war damage, inadequate maintenance, and the limited availability of spare parts, equipment, new technology, and financing. Considerable looting after Operation Iraqi Freedom and continued attacks on crude and refined product pipelines have contributed to Iraq's reduced crude oil production and export capacities.

Iraq's crude oil reserves, estimated at a total of 115 billion barrels, are the third largest in the world. However, Iraq's ability to extract these reserves has varied widely over time and has been significantly affected by war. Figure 2 shows Iraq's daily average crude oil production levels annually from 1970 through 2005.[4] In the 5 years preceding the 2003 invasion, crude oil production averaged 2.3 mbpd.

Iraq's crude oil production reached it highest annual average, 3.5 mbpd, in 1979. In September 1980, Iraq invaded Iran and production levels plummeted. Although the Iran-Iraq War continued until 1988, production levels grew steadily after 1983, peaking at 2.9 million barrels per day in 1989. The following year, Iraq invaded Kuwait, which began the Gulf War. In January 1991, the United States and coalition partners began a counteroffensive (Operation Desert Storm). Crude oil production once again dropped precipitously and remained relatively low from 1990 to 1996, while Iraq was under UN sanctions. Under the UN Oil for Food program, Iraqi crude oil production began to rebound, peaking at an annual average of 2.6 mbpd in 2000. In March 2003, the United States and coalition partners invaded Iraq. Crude oil production dropped again to a low of about 1.3 million barrels per day (annual average) in 2003 but then rebounded relatively quickly.

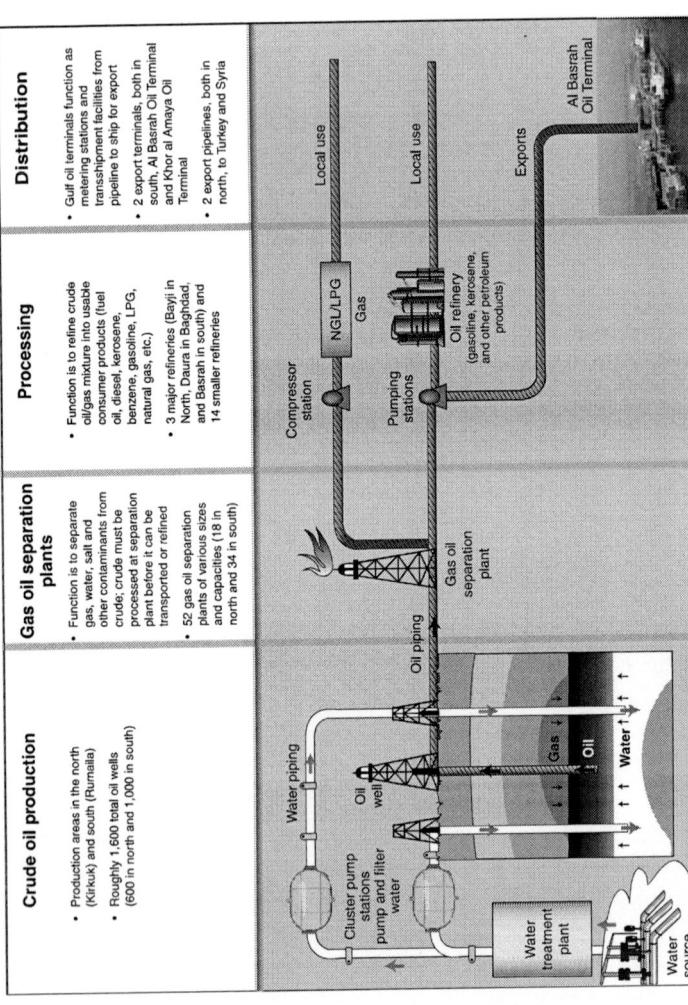

Figure 1. Overview of Oil Network.

Sources: GAO; photo (U.S. Army Corps of Engineers); clip art (Map Resources).

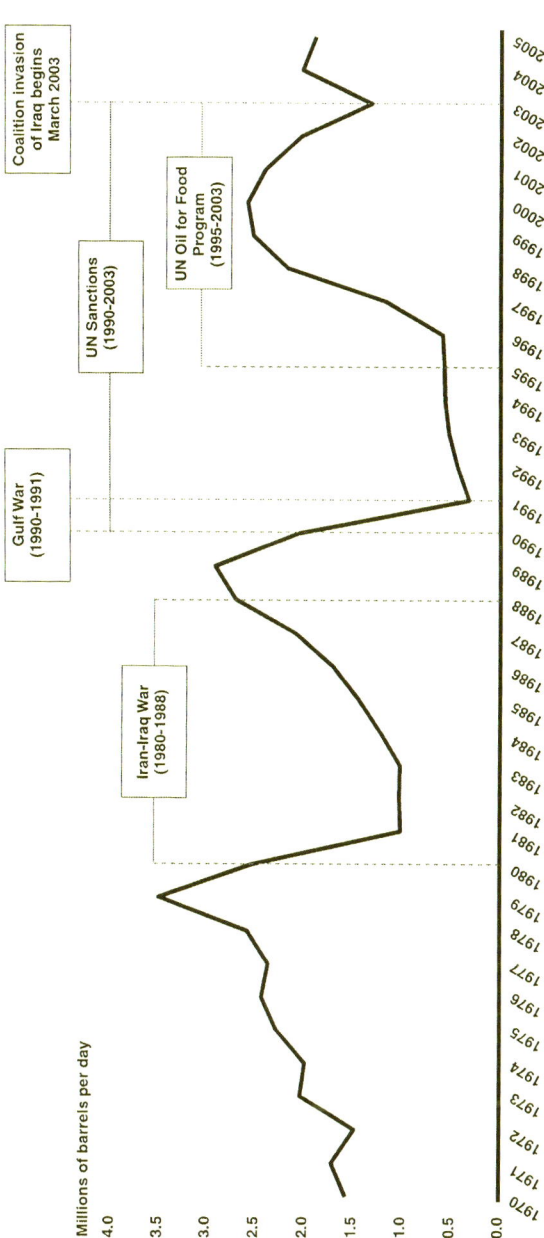

Figure 2. Crude Oil Production Levels in Iraq, 1970-2005.

Source: GAO analysis of data from the U.S. Department of Energy, Energy Information Administration.

Note: Data on Iraq's crude oil production include lease condensates. Lease condensates are hydrocarbons that are gaseous underground but become liquid above ground and are subsequently commingled with the crude oil. Therefore, the actual values for Iraq's crude oil production would be slightly lower if lease condensates were excluded from the data.

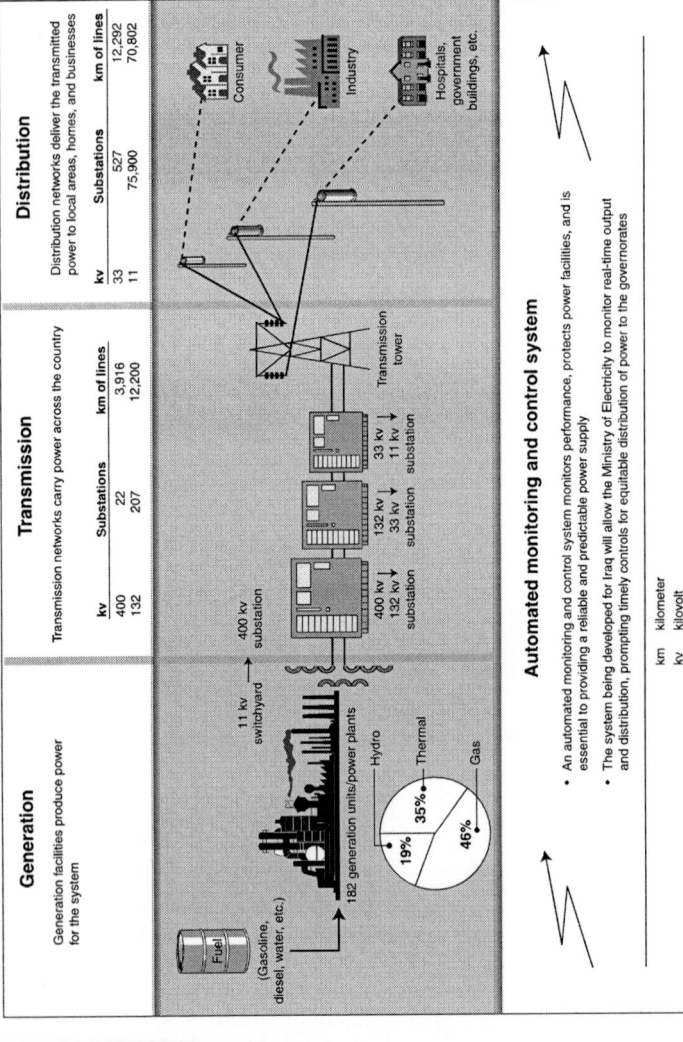

Figure 3. Overview of Electricity Network.

Sources: GAO analysis of U.S. Army Corps of Engineers (USACE) and State Department's Iraq Reconstruction Management Office data; USACE and Nova Development (clip art).

Iraq's electricity infrastructure consists of a network of (1) generation facilities that produce power; (2) transmission stations and lines that transmit power from power stations to distribution networks; (3) distribution stations and lines that move power to the end users; and (4) an automated monitoring and control system, under development, which is a centralized communications and control system designed to monitor system performance and control equitable distribution of power (see fig. 3). According to senior U.S. agency officials, Iraq's electricity infrastructure was in worse condition following the 2003 conflict than initially anticipated or reported in the 2003 UN/World Bank Needs Assessment.[5] The report noted that the severe degradation of Iraq's generating capacity—from about 5,100 megawatts in 1990 to about 2,300 megawatts after the 1991 Gulf War—was largely due to war damage to generation stations. Although the report notes that production was restored to about 4,500 megawatts before the 2003 conflict, U.S. officials said that Iraq's electrical infrastructure had experienced significant deterioration due to the war and years of neglect under Saddam's regime.

From May 2003 through June 2004, the Coalition Provisional Authority (CPA), led by the United States and the United Kingdom, was the UN-recognized authority responsible for the temporary governance of Iraq and for overseeing, directing, and coordinating reconstruction efforts. In 2003, the CPA committed to improving the oil sector in Iraq and stated that its activities were to include repairing and restoring the oil infrastructure to pre-war levels, among other goals. The CPA also committed to improving the electricity sector in Iraq and stated that CPA partners would provide reliable, stable, and predictable power to the people and enterprises of Iraq by rebuilding generation, transmission, and distribution networks, as well as system control and communications; by establishing a sound regulatory framework for the electricity sector; and by ensuring improved security for infrastructure.

In developing its reconstruction plan, the CPA made several assumptions that never materialized.[6] The first and most critical assumption was that there would be a permissive security environment to conduct reconstruction activities. Second, the CPA assumed that U.S.-funded reconstruction activities would help restore Iraq's essential services—including oil production and electricity generation—to prewar levels. The CPA assumed that the Iraqi government and the international community would help finance Iraq's developmental needs and that Iraqi oil revenues could help pay for reconstruction costs. The CPA estimated that Iraq's oil production would increase to about 2.8 to 3 mbpd by the end of 2004.

Additionally, it assumed that because of its expertise, the United States would focus its resources on long-term reconstruction projects.

With the establishment of Iraq's interim government in June 2004, the CPA's responsibilities were transferred to the Iraqi government or to U.S. agencies. Since then, the Department of State has been responsible for overseeing U.S. efforts to rebuild Iraq. The Project and Contracting Office (PCO), a temporary DOD organization, was tasked with providing acquisition and project management support. In December 2005, DOD merged the PCO with the U.S. Army Corps of Engineers Gulf Region Division, which supervises DOD reconstruction activities in Iraq. Additionally, the State Department's Iraq Reconstruction and Management Office (IRMO) has been responsible for strategic planning and for prioritizing requirements, monitoring spending, and coordinating with the military commander. According to U.S. officials, IRMO is scheduled to sunset on May 10, 2007, and is to be replaced by the Office of Transition and Assistance Coordination. IRMO advisors to the ministries will continue their work under the new office or other relevant U.S. offices. As of March 2007, IRMO had assigned the Ministries of Oil and Electricity 10 and 18 advisors, respectively. Additionally, USAID awarded and managed its own contracts for electricity under its Bechtel contracts which, as of April 2007, were being closed out. USAID continues to award its own contracts, which are now generally associated with economic assistance, governance, capacity building, and its community stabilization program.

THE UNITED STATES, IRAQ, AND DONORS HAVE FUNDED RECONSTRUCTION OF IRAQ'S OIL AND ELECTRICITY SECTORS, BUT IRAQ'S FUTURE NEEDS ARE SIGNIFICANT AND SOURCES OF FUNDING UNCERTAIN

For fiscal years 2003 through 2006, the United States made available about $7.4 billion, obligated about $7.1 billion, and spent about $5.1 billion in U.S. funds to rebuild Iraq's oil and electricity sectors. The U.S. government also has spent about $3.8 billion in Iraqi funds, as of December 31, 2005, for reconstruction activities, including CPA contracts administered by U.S. agencies. However, according to various estimates, each sector will need billions more to achieve its rebuilding goals. Since the majority of the U.S. funds have been spent, the Iraqi government and international donor community represent important sources of potential future funding. However, these sources of funding remain uncertain. The Ministries of Oil and Electricity have encountered difficulties spending their existing capital improvements budgets, and their future budgets are subject to the volatility of oil revenues. The Iraqi government has not made full use of potential international contributions, and future donor funding is also uncertain.

UNITED STATES HAS SPENT BILLIONS IN U.S. AND IRAQI FUNDS TO SUPPORT RECONSTRUCTION

The United States has spent the majority (about 70 percent) of the $7.4 billion in funds it made available to reconstruct the electricity and oil sectors. Additionally, the United States spent $3.8 billion in Iraqi funds on oil and electricity sector reconstruction activities.

United States Has Spent Billions to Support Reconstruction of Iraq's Oil and Electricity Sectors

U.S. agencies have spent about $5.1 billion of the $7.4 billion made available for the oil and electricity sectors, as of September 30, 2006 (see table 1). For the oil sector, U.S. agency efforts focused primarily on production and exportation, and to a lesser extent on the rehabilitation of refinery and gas facilities. For the electricity sector, U.S. agency efforts generally focused on restoring or constructing generation, transmission, distribution, and automated monitoring and control system projects.

Funding for electricity included about $164 million for the Commander's Emergency Response Program (CERP) during fiscal years 2004 to 2006.[7] Additional funding for CERP has been provided under the Department of Defense 2007 Appropriations Act,[8] which makes available up to $500 million for CERP in Iraq and Afghanistan. The fiscal year 2007 CERP allocation for electricity projects in Iraq was not available as of March 2007, according to Army and DOD officials. CERP provides coalition military commanders with a tool to rapidly respond to urgent humanitarian relief and reconstruction needs in their geographic area of responsibility.

Table 1. Status of Appropriation Funds Apportioned to the Oil and Electricity Sectors, as of September 30, 2006

Dollars in millions

Source of funds[a]/authority	Agency/program	Apportioned	Obligated	Expended
Oil sector				
2003 Iraq Relief and Reconstruction Fund (IRRF 1)/ P.L. 108-11	DOD/Restore Iraq Oil	$166.0	$166.0	$166.0
The Natural Resources Risk Remediation Fund (NRRRF)/ P.L. 108-11	DOD/ Restore Iraq Oil	802.0	800.6	797.7
2004 Iraq Relief and Reconstruction Fund (IRRF 2)/ P.L. 108-106	DOD	1,724.7	1,604.6	1,163.0
Total oil		**$2,692.7**	**$2,571.2**	**$2,126.7**
Electricity sector				
IRRF 1/ P.L. 108-11	DOD / Restore Iraq Electricity	$300.0	$299.9	$299.9
	USAID/ Restore Critical Infrastructure[b]	Not available	Not available	Not available
IRRF 2/ P.L. 108-106	DOD	3,403.0	3,260.1	1,966.0
	USAID	836.6	836.6	741.9
Commander's Emergency Response Program for FY2004-06 (CERP)[c]/P.L. 108-106, P.L. 108-287, P.L. 109-13, P.L.109-148, P.L. 109-234	DOD	163.6	163.6	Not available
Total electricity		**$4,703.2**	**$4,560.2**	**$3,007.8**
Total		**$7,395.9**	**$7,131.4**	**$5,134.5**

Sources: DOD, USAID, and State Department's Iraq Reconstruction and Management Office data.

[a] According to the DOD comptroller's office, although some military construction funds also were used for reconstruction efforts, they were not apportioned by sector. Further, about $1.1 million of Operations and Maintenance, Army funds were initially apportioned, obligated, and expended for electricity projects. Also, $285 million in emergency supplemental funds were made available for cross-sector operations and maintenance sustainment of key U.S.-funded infrastructure.

[b] According to USAID, IRRF 1 funds under its Restore Critical Infrastructure program were not apportioned by sector. Under this cross-sector reconstruction program, about $995 million was apportioned and obligated and about $939 million was disbursed as of September 30, 2006.

[c] DOD and its executing agency, the U.S. Army, were unable to provide complete data for U.S.-appropriated CERP funding for the electricity sector. Under CERP, DOD commits funds to projects once identified. DOD does not track sector commitments after the year of the corresponding appropriation. Thus, any updated data on amounts apportioned, obligated, or expended by sector are not available and not reflected in this table.

U.S. Agencies Spent Billions in Iraqi Funds on Oil and Electricity Reconstruction Activities

U.S. agencies spent about $3.8 billion in Iraqi funds in the oil and electricity sectors, as of December 31, 2005 (see table 2). These funds came from vested and seized assets from the previous Iraqi regime[9] and oil revenues in the Development Fund for Iraq (DFI).[10] The CPA spent vested and seized assets on reconstruction activities, including projects, ministry operations, and liquefied petroleum gas. DFI funds consist of oil proceeds, UN Oil for Food program surplus funds, and returned Iraqi government and regime financial assets; DFI funds continue to be used for contracts administered by U.S. agencies.[11] According to International Advisory Monitoring Board of the DFI documents, the Iraqi government granted U.S. agencies limited authority to administer outstanding contracts entered into by the former CPA that required subsequent payments after June 28, 2004.[12] This authority expired on December 31, 2006. In March 2007, the U.S. government requested that the Iraqi government extend the authority to December 31, 2007.

Table 2. Iraqi Funds Used for U.S. Oil and Electricity Sectors, as of December 31, 2005

Oil sector	Disbursed
Development Fund for Iraq	$2,709,267,000
Seized funds	95,000,000
Vested funds	7,000
Total for oil sector	**$2,804,274,000**
Electricity sector	
Development Fund for Iraq	$1,014,674,000
Seized funds	2,975,000
Vested funds	12,925,000
Total for electricity sector	**$1,030,574,000**
Total	**$3,834,848,000**

Source: GAO analysis based on data from CPA, DOD, and the independently audited DFI Statement of Cash Receipts and Payments, as reported to the Iraqi government and the International Advisory Monitoring Board.

Notes: Current independent auditor's data for the DFI are as of December 31, 2005. The results of the independent audit of 2006 funds have not yet been released. Sector activities include the DFI purchase of about $2,241 million in petroleum products and $50 million in generated electricity from other countries. Disbursed figures do not include DFI funds provided through the CERP and other small programs, which also support reconstruction activities.

Billions of Additional Dollars Needed to Rebuild Oil and Electricity Sectors, but Future Iraqi Funding Is Uncertain

According to the Iraqi government, the UN, and the U.S. government, Iraq's Ministries of Oil and Electricity will need billions of additional dollars to rebuild, maintain, and secure Iraq's oil and electricity infrastructure and achieve production goals. However, the Iraqi government has had difficulty fully expending its funds already allocated to the electricity and oil sectors. In addition, the amount of Iraq's future budget revenues is uncertain due to Iraq's dependency on potentially volatile oil revenues. According to the World Bank, volatility in oil export revenues could hamper the prospects for Iraq's reconstruction program because volatility increases uncertainty, leads to wasteful public investment in times of boom, and depresses investment when oil prices are low.

Billions of Additional Dollars Are Needed to Fund the Rebuilding of the Oil and Electricity Sectors

The Iraqi government, the UN, and the U.S. government estimate that Iraq will need billions of additional dollars to continue rebuilding, maintaining, and securing Iraq's oil and electricity infrastructure. For example, the Ministry of Electricity's 2006-2015 Electricity Master Plan estimates that $27 billion will be needed to reach its goal of providing reliable electricity across Iraq by 2015. The goals of the Master Plan are to (1) rehabilitate the existing power generation plants and transmission and distribution networks, (2) increase generation capacity, (3) provide a secure power supply to all consumers, (4) build the capacity of ministry staff to implement the plan, and (5) connect Iraq's grid with neighboring countries. The achievement of these goals is intended to ensure that Iraq meets its future demands for electricity, which are expected to double by 2015. Under the master plan, the demand for electricity would be first met in 2009.

For the oil sector, a comparable estimate of future rebuilding needs has not been developed. According to DOD, the investment in Iraq's oil sector is "woefully short" of the absolute minimum required to sustain current production and additional foreign and private investment is needed for the development of Iraq's energy sector. Moreover, U.S. officials and industry experts have stated that Iraq would need an estimated $20 billion to $30 billion over the next several years to reach and sustain a crude oil production capacity of 5 million barrels per day. This production goal is below the level identified in the Iraqi 2005-2007 National Development Strategy--at least 6 million barrels per day by 2015.

Iraq's Oil and Electricity Ministries Have Not Spent Existing Funds for Infrastructure Improvements and Future Iraqi Funds Are Uncertain

The Iraqi government has not fully spent the capital project funds already allocated to the electricity and oil sectors in Iraq's 2006 budget. According to U.S. and foreign officials, the Ministries of Oil and Electricity have encountered difficulties spending these budgets because of government weaknesses in budgeting, procurement, and financial management. While Iraq's inability to expend its capital budgets may not directly affect U.S.-funded projects, U.S. investments alone are not adequate for the full reconstruction and expansion of the oil sector. Therefore, Iraq's continued difficulties in spending its capital budgets could hamper efforts to attain its current reconstruction goals. In 2006, Iraq planned to spend more than $3.5 billion and $767 million for capital projects in the oil and electricity sectors, respectively. These amounts accounted for about 98 percent of the Ministry of Oil's total budget ($3.6 billion) and 91 percent of the Ministry of Electricity's total budget ($840 million) that year (see fig. 4). However, as shown in figure 4, only 3 percent of oil sector capital project funds had been spent and about 35 percent had been spent in the electricity sector, as of November 2006.[13] We are examining U.S. efforts to work with the Iraqi ministries regarding ministry budget execution issues in greater detail in another ongoing review.

According to U.S. officials, Iraq lacks the clearly defined and consistently applied budget and procurement rules needed to effectively implement capital projects. For example, the Iraqi ministries are guided by complex laws and regulations, including those implemented under Saddam Hussein, the CPA, and the current government. According to State officials, the lack of agreed-upon procurement and budgeting rules causes confusion among ministry officials and creates opportunities for corruption and mismanagement. Additionally, according to State and DOD, personnel turnover within ministries, fear of corruption charges, and an onerous contract approval process[14] have caused delays in contract approval and capital improvement expenditures. Moreover, according to IRMO officials, the Ministry of Oil faces continued difficulties in obtaining Ministry of Finance and other official approvals for meeting its operations and maintenance expenses, procuring needed replacement items, and funding critical emergency repairs and new projects. According to the World Bank, Iraq's procurement procedures and practices are weak and lack effective bid protest mechanisms and transparency on final contract awards, among other problems.[15]

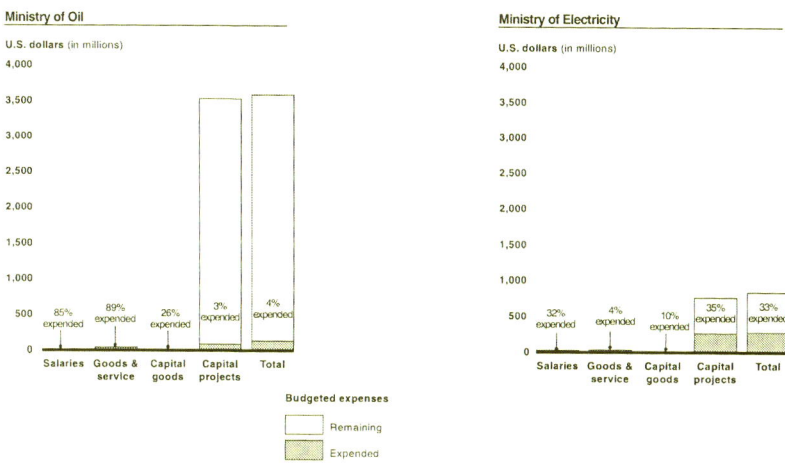

Source: GAO analysis of Iraq's budget, provided by U.S. Treasury.

Figure 4. 2006 Ministry of Oil and Electricity Spending by Major Category, as of November 2006

In response to the problems that Iraq is experiencing in spending its budgets, the United States is considering some additional options for facilitating Iraq's expenditure of these capital project funds. In April 2007, the State Department authorized DOD, under section 607 of the Foreign Assistance Act of 1961, as amended (Public Law 87-195), to provide commodities and services to the Iraqi government on an advance-of-funds basis. DOD has identified areas in which the government of Iraq may seek its help, including the procurement of fuel, electrical power generation, water projects, and security upgrades. According to DOD officials, Iraq must formally request this assistance before such a mechanism could be utilized. These officials indicated that no additional details could be provided since the details are still being worked out with the Iraqi government. We are monitoring the use of this mechanism as part of our ongoing work on Iraq and will report separately on this issue at a later date.

In addition to the challenges in spending its current budgets, Iraq's future budget revenues are subject to some uncertainty due to Iraq's high dependence on crude oil export revenues. Oil exports are estimated to be $31 billion in 2007, accounting for more than 90 percent of the government revenue budgeted for fiscal year 2007. Iraq's oil revenues are determined by the amount of crude oil Iraq exports and its price. While crude oil exports have been relatively stable, the price of Iraqi crude oil has been much more variable. The price of Iraqi crude oil rose from a little over $30 per barrel in 2004 to above $60 per barrel during the summer of 2006 and then fell again to about $50 per barrel at the end of 2006. In the first quarter of 2007, prices were once again rising. These fluctuations, which are dependent on a wide range of geopolitical and economic factors, make it hard to reliably forecast government revenues. According to the World Bank, volatility of oil export revenues may hamper the prospects of Iraq's reconstruction program, since volatility increases uncertainty, leads to wasteful public investment in times of boom, and depresses investment when oil prices are low.

IRAQ HAS NOT MADE FULL USE OF INTERNATIONAL FINANCIAL ASSISTANCE AND PROSPECTS FOR FUTURE FUNDING ARE UNCERTAIN

International donor funds represent an important source of potential funding. Potential funding for Iraq's oil and electricity sectors is available in the form of bilateral grants and loans, multilateral grants through the International Reconstruction Fund Facility for Iraq,[16] and multilateral loans through the World Bank, the International Monetary Fund, or the Islam Development Bank.

However, as of March 2007, donors had not provided any grants to the oil sector, and the Iraqi government had not taken advantage of available loans for either the oil or the electricity sector. Some grants have been provided to the electricity sector. As of March 2007, of the more than $14 billion in overall non-U.S. assistance pledged for Iraq, about $580 million had been committed and $441 million spent on grants for electricity projects.[17] According to U.S. and donor officials, no donor funding was provided to the oil sector because of an expectation that sufficient funds would be provided through Iraq's oil revenues and private investors.

Although loans have been offered to the Iraqi government, it has not accessed this source of financing to date due, in part, to its concerns about adding to its considerable debt burden. For example, Japan has announced its intention to provide yen loans to the Iraqi government for several projects totaling about $1.1 billion for the electricity and oil sectors.[18] However, according to the State Department, as of January 2007, the Iraqi government had not reached agreement with Japan on the terms of the loans. As we previously reported, Iraq has significant foreign debt remaining from the Saddam Hussein regime, which presents financial challenges for Iraq's reconstruction and economic development.[19]

Moreover, it is unclear to what extent the International Compact with Iraq will serve as a viable mechanism to obtain additional donor support for Iraq. Initiated by Iraq and the UN in July 2006, the compact is intended to promote wider international engagement in the redevelopment of Iraq and to secure additional financial assistance to support major investments in sectors such as oil, electricity, and agriculture. Under the compact, Iraq would undertake economic, political, and security reforms to receive additional donor support. According to the State Department, the compact was signed and formally launched on May 3, 2007. According to the State Department, the international community will be able to contribute under the compact in a number of ways including debt reduction or forgiveness, pledging new financial and technical assistance, and expediting the disbursement of previously pledged assistance. However, the extent to which the compact can be expeditiously implemented and stimulate international assistance remains uncertain.

IRAQ'S OIL AND ELECTRICITY PRODUCTION GOALS HAVE NOT BEEN MET, OIL PRODUCTION FIGURES MAY BE OVERSTATED, AND IRAQ FACES DIFFICULTIES SUSTAINING INFRASTRUCTURE

Oil and electricity production have consistently fallen below U.S. program goals, undermining U.S. and Iraqi government efforts to improve essential services. Production levels may be overstated and measuring them precisely is challenging due to limited metering and poor security. Further, the demand for electricity significantly outstripped the available supply in 2006. To help improve Iraq's oil and electricity production, the United States has funded a variety of projects. However, some of these projects, which could provide additional production capacity, have faced significant delays undermining efforts to meet U.S. program goals. Additionally, Iraq has difficulty sustaining U.S.-funded infrastructure projects once they are completed.

U.S. OIL AND ELECTRICITY GOALS HAVE NOT BEEN MET

Key reconstruction goals for Iraq's oil and electricity sectors, including those for crude oil production and exports, refined fuel production capacity and stock levels, and electricity production, among others, have not been met. In addition, Iraq's oil production levels may be overstated given problems with inadequate metering. Department of Energy, for example, reports Iraq's oil production at 100,000 to 300,000 barrels per day less than what the State Department reports. Production goals for electricity are outweighed by increasing demand, which is difficult to measure and continues to significantly exceed supply.

Iraq's Crude Oil Production and Exports Have Not Met Goals

Key reconstruction goals for Iraq's oil sector have yet to be achieved (see fig. 5). Key U.S. goals for the oil sectors are to reach an average crude oil production capacity of 3 million barrels per day (mbpd) and crude oil export levels of 2.2 mbpd, and increase production capacity and stock levels of refined fuels.[20] However, in 2006, actual crude oil production and exports averaged, respectively, about 2.1 mbpd and 1.5 mbpd, and refined fuel production and stock levels did not meet their goals.

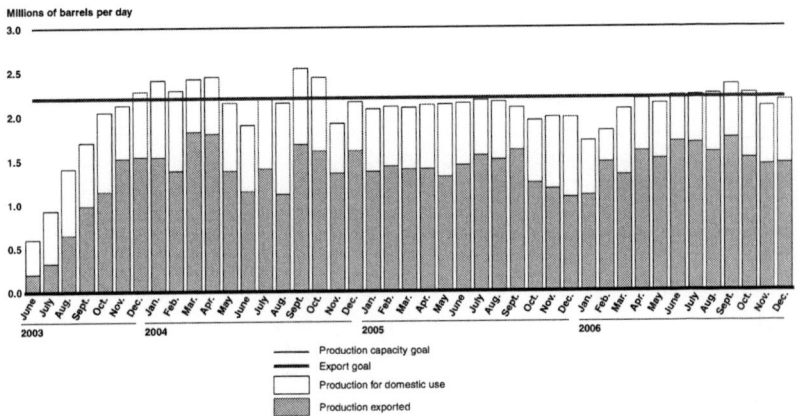

Source: Iraq Ministry of Oil estimates collected by State Department's Iraq Reconstruction and Management Office.

Figure 5. Iraqi Reported Crude Oil Production, Exports, and U.S. Goals, June 2003 through December 2006

In August 2003, the CPA established a U.S. program goal to increase crude oil production to about 1.3 mbpd. The CPA increased this goal every 2 to 3 months until July 2004, when the goal became to increase crude oil production capacity to 3.0 mbpd (shown in fig. 5). State also set an eventual crude oil production goal of 2.8 mbpd in March 2006. Production capacity differs from actual production. Production capacity is the maximum amount of production a country can maintain over a period of time. For example, EIA has defined production capacity as the maximum amount of production that (1) could be brought online within 30 days and (2) sustained for at least 90 days. Since Iraq has been trying to increase its production of crude oil, we use actual production as an indicator of Iraq's production capacity in this report.

Table 3. U.S. Goals and 2006 Averages for Iraq Oil Sector

Metric	U.S. Goal	2006 Average	2007 Average, as of February 28, 2007
Crude oil production (million barrels per day)	3.0 mbpd (capacity)	2.1 mbpd	1.9 mbpd
Crude oil exports (million barrels per day)	2.2 mbpd	1.5 mbpd	1.4 mbpd
Natural gas production (million standard cubic feet per day)	800 mscfd (capacity)	730 mscfd	N/A
LPG production (tons per day)	3,000 tpd (capacity)	1,709 tpd	N/A
National stock levels of refined fuels: diesel, kerosene, gasoline, LPG	15 days for each fuel	Diesel: 5 days Kerosene: 10 days Gasoline: 3 days LPG: 5 days	Diesel: 3.5 days Kerosene: 4 days Gasoline: 2 days LPG: 1 days

Source: GAO analysis based on State Department and Defense Department data.

Besides production and export of crude oil, the CPA also established goals for the production of natural gas and liquefied petroleum gas (LPG), as well as the national stocks of refined petroleum products (such as gasoline) that are used to generate energy by consumers and businesses.[21] These CPA goals were to increase production capacity of natural gas to 800 million standard cubic feet per day (mscfd); increase production capacity of LPG to 3,000 tons per day (tpd); and meet demand for benzene (gasoline), diesel, kerosene, and LPG by building and maintaining their stock levels at a 15-day supply. However, the 2006 averages did not meet these goals, as shown in table 3. To increase the stocks of petroleum products and the availability of these products to consumers, Iraq legalized the importation of petroleum products by private companies to supplement its own production and state-owned company imports. For 2006, the IMF estimated that Iraq's state-owned companies imported about $2.6 billion of petroleum products. At the recommendation of the IMF, the Iraqi government has also been reducing subsidies for refined oil products, which raises the prices consumers pay. In the past, refined oil products in Iraq have been highly subsidized, which leads to increased demand. Reduction in domestic demand for refined oil products would allow additional crude oil to be exported for revenue rather than refined in Iraq.

Iraq's Crude Oil Production May Be Overstated

Iraq's crude oil production statistics may overstate Iraq's oil production. We compared IRMO's statistics to those published by the EIA, which are based on alternate sources.[22] Part of EIA's mission is to produce and disseminate statistics on worldwide energy production and use. While these two data sets follow similar trend lines, EIA's series is generally 100,000 to 300,000 barrels per day lower than IRMO's statistics. At an average price of $50 per barrel, this is a discrepancy of $5 to $15 million per day. Figure 6 shows these two data sets over the time period (June 2003 to December 2006) for which data were available for both. According to EIA, several factors may account for the discrepancy. One known factor is the lack of storage facilities for crude oil in Iraq. With limited storage, the crude oil in Kirkuk that cannot be either processed by refineries or exported (both of which have been hampered by limited refinery capacity and sabotage to oil pipelines) is re-injected into the ground after the natural gas liquids are removed.[23] Another factor affecting the discrepancy may be differences in the frequency and timing of the data. IRMO's data is reported daily in real time, while EIA produces monthly data that has been reviewed and corroborated from several sources. This lag in reporting and longer time period may allow analysts to address inconsistencies such a double-counting and reinjection. In addition, IRMO regularly reports on sabotage and interdictions to crude oil pipelines and other disruptions in the crude oil production process. Also, under Saddam Hussein, Iraq had a history of diverting crude oil production to circumvent UN sanctions. Therefore, it is possible that corruption, theft, and sabotage may also be factors in the discrepancy.

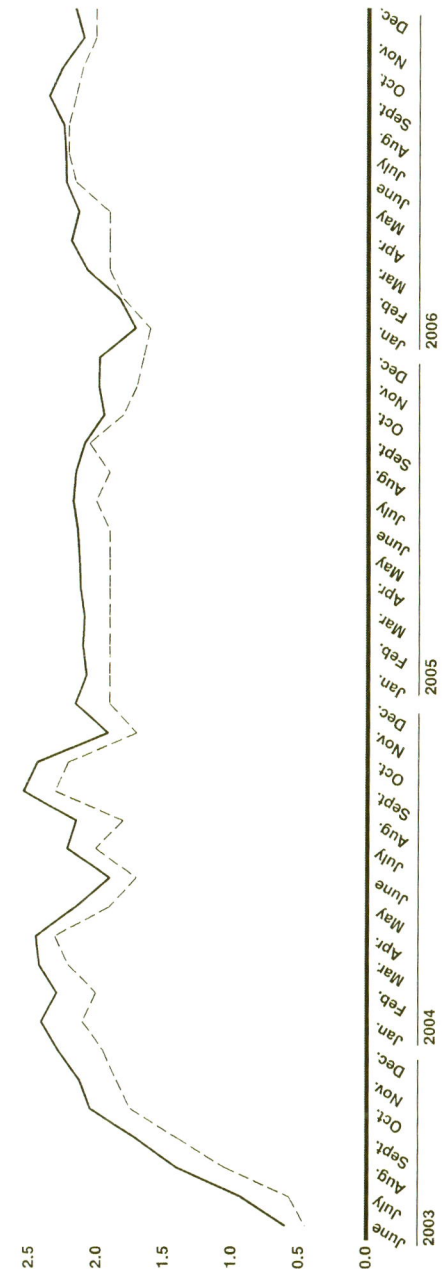

Figure 6. Comparison of IRMO and EIA Data on Iraq's Crude Oil Production.

Source: Iraq Ministry of Oil estimates collected by State Department's Iraq Reconstruction and Management Office and U.S. Energy Information Administration data.

However, identifying, measuring, and resolving these factors without precise metering is more difficult. Without knowing precisely how much crude oil is being produced and transported along the extensive network of pipelines, refineries, and export facilities, the government is limited in its ability to identify potential weaknesses in the system. According to an IRMO oil advisor, meters are in place at many locations but are not usable in many instances due to the difficulties in obtaining needed replacements and spare parts. Without comprehensive metering, crude oil production must be estimated using less precise means, such as estimating flow through pipelines, and relying on reports from on-site personnel rather than an automated system that could be verified.

An improved metering system has been a U.S. and international donor priority since early 2004 but has faced delays in its implementation. In 1996, the UN first cited the lack of oil metering when Iraq was under UN sanctions. In 2004, the International Advisory and Monitoring Board (IAMB) for the Development Fund for Iraq recommended the expeditious installation of metering equipment, in accordance with standard oil industry practices. According to the IAMB, in June 2004, the CPA had approved a budget to replace, repair, and calibrate the metering system on Iraq's oil pipeline network and to contract the metering of Iraq's oil resources. However, the oil metering contract was not completed due to security and technical issues. In June 2006, the IAMB reported that the Iraqi government had entered into an agreement with Shell Oil Company to serve as a consultant for the Ministry of Oil on metering and calibrating that would include the establishment, within the next 2 years, of a measuring system for the flow of oil, gas, and related products within Iraq and export and import operations. The U.S. government is assisting in this effort by rebuilding one component of the metering system in the Al-Basrah oil port—Iraq's major export terminal—and expects the project to be complete by June 2007.

The collection of reliable statistics is also complicated by the challenging security environment and the numerous entities involved in compiling countrywide production data (from wellheads, through pipelines, to final uses in refineries or exports). Iraq's current oil operations are divided among several state-owned oil companies that have responsibility for operations. Therefore, IRMO, which is responsible for producing aggregate data on Iraq's production levels, must collect daily reports from the Ministry of Oil and each of the state-owned enterprises and combine them into a countrywide total. IRMO tries to identify and correct any anomalies and monitors the data on a daily basis.

Iraq's Electricity Production Has Not Met Goals or Increasing Demand

Key reconstruction goals for Iraq's electricity sector have yet to be achieved and demand continues to outpace supply. In 2004, the CPA established a U.S. program goal to improve peak generation capacity to 6,000 mw per day by the end of June 2004. However, by the end of 2006, peak generation capacity for the year averaged only 4,280 mw per day and the goal remained at 6,000 mw per day.[24] In March 2006, the State Department also set a goal to achieve 12 hours of power per day both in Baghdad and nationwide. In 2006, hours of power per day nationwide averaged 11 and only 5.9 in Baghdad. This was well below the goal, as shown in table 4.

Table 4. U.S. Goals and 2006 Averages for Iraq Electricity Sector

Metric	U.S. Goal	2006 Average	2007 Average, as of February 28, 2007
Peak generation	6,000 mw (capacity)	4,280 mw	3,803 mw
Average daily generation	110,000 mwh	94,665 mwh	84,796 mwh
Hours of power nationwide and in Baghdad	12 hours (hrs)	Nationwide: 11.0 hrs Baghdad: 5.9 hrs	Nationwide: 8.6 hrs Baghdad: 5.1 hrs

Source: GAO analysis based on State Department's Iraq Reconstruction and Management Office data.

The State Department also aimed to reach 110,000 megawatt hours (mwh)[25] of average daily generation in 2005 and 2006 (see fig. 7).[26] However, average daily peak generation for 2006 reached 94,665 mwh, also below the goal.

Further, demand has continued to exceed supply throughout 2005 and 2006, reaching a peak demand of 200,744 mwh for the week of August 20, 2006 (see fig. 7).[27] For 2006, demand averaged 8,180 mw exceeding the average peak electricity supply of 4,280 mw by about 3,950 mw. According to U.S. agency officials, demand for electricity has been stimulated by a growing economy and a surge in consumer purchases of appliances and electronics. In addition, electricity is subsidized in Iraq, which leads to increased demand. If the Ministry of Electricity's master plan for 2006 to 2015 to rehabilitate and expand the national grid is implemented, the ministry estimates that Iraq will be able to meet its projected demand for

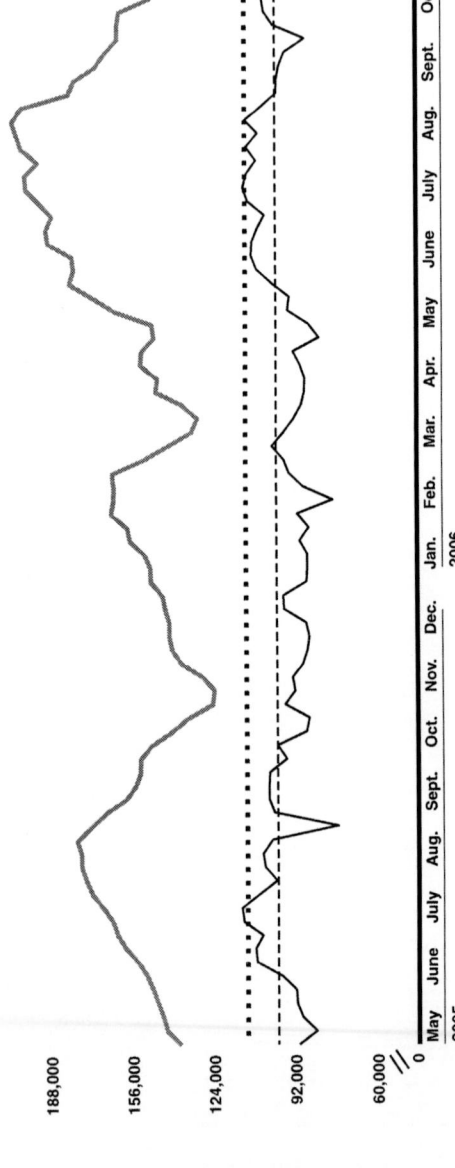

Figure 7. Peak Electricity Generation and Demand in Iraq, May 2005 to December 2006.

Source: Iraq Ministry of Electricity estimates collected by Department of State.

electricity in 2009. However, these projections assume a stable supply of fuel for electricity generation, which has been lacking in the past due to poor coordination between the ministries. The projections are also based on the assumption that the Ministry of Electricity will be able to mobilize funding for projects listed in the Master Plan and properly manage, maintain, and operate its electricity infrastructure.

Moreover, the widespread use of off-grid electrical generators to meet unmet demand from the national grid makes it difficult to measure total demand and the progress made toward meeting that demand. According to USAID, it is estimated that neighborhood generators provide an additional 2,000 mw per day. However, electricity usage from neighborhood generators cannot be accurately measured, according to IRMO officials. Likewise, when measuring production, it is important to note that the U.S. and Iraqi goal is to supply homes from the national grid rather than from the inefficient use of small household and neighborhood generators. Thus, production from those generators would not be considered in meeting production targets.

MOST MAJOR U.S. PROJECTS COMPLETED, BUT IRAQ FACES DIFFICULTIES SUSTAINING U.S. REBUILDING EFFORTS

The United States has completed the majority of its planned projects in the oil and electricity sectors, funded through the fiscal year 2004 Iraq Relief and Reconstruction Fund (IRRF 2).[28] Some of the remaining oil projects that could help meet U.S. program goals have faced delays due to a variety of factors such as scope changes during engineering and design phase, delays in mobilizing subcontractors, the poor condition of equipment, and the insecure working environment. Moreover, Iraq has experienced difficulty in sustaining the infrastructure rehabilitated with U.S. funds.

As of March 2007, the Army Corps of Engineers GRD reported that 88 percent of its work on planned IRRF oil projects had been completed and that it expected to complete all projects by May 2007. In the electricity sector, GRD reported in March 2007 that 73 percent of its work on planned IRRF 2 electricity projects had been completed and that it expected to complete its other projects by April 2008. According to DOD, this remaining work will enable Iraq to increase crude oil production capacity by 0.5 million barrels per day and petroleum gas production by 1,800 tons per day.

In the electricity sector, GRD's IRRF 2 projects, once completed, will provide a total of 1,879 mw of potential generation capacity, which could serve an estimated 1.7 million homes. According to GRD reporting, all U.S. agency projects (GRD and USAID) are expected to add or restore a total of 2,555 mw when completed.

Some key U.S. projects are likely to face continued schedule slippage or transfer to state oil companies before they are completed, thereby undermining efforts to reach U.S. program goals. The following illustrates the issues facing two U.S. projects.

- *Qarmat Ali Pressure Maintenance.* GRD has reported that the United States has committed about $105 million for three projects ($41 million, $27 million, and $37 million) to restore the Qarmat Ali water reinjection and treatment plant to create and maintain sufficient oil field pressure in the Rumailah oil field, one of the largest southern oil fields. According to GRD engineers, the final completed project could result in an increase in crude oil production capacity of 200,000 barrels per day. As of April 2007, 56 percent of the work of the last project has been completed. The Special Inspector General for Iraq Reconstruction (SIGIR) has reported schedule slippage caused by the poor condition of the original pumps, motors, and valves. According to IRMO oil officials, due to lack of funding for the administrative task orders that cover life support and security, the construction phase of this project was de-scoped and responsibility was transferred to Iraq's South Oil Company. Furthermore, IRMO oil officials question whether the project will increase production; they contend that the water injection will at best reduce the decline of the oil field's production.
- *South Well Workover.* According to IRMO officials, in the refurbishing of wells throughout Basrah, delays have been caused by poor contractor performance, bad equipment, lack of spares, and subcontract disputes. GRD has reported that the United States has committed $93 million for this project, including the workover of 30 wells in the Rumaila fields and completion and replacement of tubing in 50 wells in West Qurna.[29] This project began in June 2006 and had an expected completion date of December 2006. As of April 2007, the work was completed and could result in an increase in crude oil production capacity of 150,000 barrels per day (50 wells at 3,000 bpd).

Further, despite capacity building efforts, Iraq has experienced difficulty in sustaining the infrastructure rehabilitated with U.S. funds. For example, the U.S. decision to install gas turbine generators without an assured supply of natural gas has resulted in increased maintenance requirements and decreased power output. Iraq's inability to sustain the new and rehabilitated oil and electricity infrastructure undermines efforts to reach U.S. program goals.

In the oil sector, U.S. agency and contractor officials report that the sector's rebuilding efforts continue to be impeded by the lack of modern technology; the lack of qualified staff and expertise at the field, plant, and ministry levels; an ineffective inventory control system for spare parts within the oil sector's 14 operating companies; and difficulties in spending budgets for equipment upgrades and replacements. The U.S. government has provided additional training and management assistance in response to these needs. According to DOD, IRRF funds were allocated toward sustainment and capacity development activities including training to state-owned oil companies on preparation of long-term service agreements and training on heavy equipment, procurement of functional and operational spare parts, and the invitation-to-bid process.

According to U.S. government reporting, the lack of an adequate operating and maintenance practices is one of many factors inhibiting a robust electrical grid in Iraq. U.S. officials also report that rebuilding of the electricity sector has been slowed by the lack of training for plant workers, inadequate spare parts, and an ineffective asset management and parts inventory system. Electricity plants are sometimes operated beyond their recommended limits and use poor-quality fuels that rapidly deteriorate parts, involve longer maintenance downtimes, and increase pollution. According to U.S. government officials, Iraq needs to develop cleaner and more reliable sources of natural gas for its generators supporting the national grid. Currently, Iraq's fuel supply does not meet demand and its quality is inconsistent. For example, of the 35 natural gas turbines the U.S. government installed in power generation plants, 16 are using diesel, crude, or heavy fuel oil due to the lack of natural gas and lighter fuels. As a result, maintenance cycles are reportedly three times as frequent and three times as costly. Poor-quality fuels also decrease the power output of the turbines by up to 50 percent and can result in equipment failure and damage, according to U.S. and Iraqi power plant officials. The U.S. government also estimates that Iraq is flaring enough natural gas to generate at least 4,000 mw of electrical power. Because of natural gas shortages, diesel has to be imported at a cost of about $1.2 billion a year, thus straining economic resources.

The U.S. government is providing some assistance to address these shortfalls. According to DOD, over $120 million of IRRF funds were allocated toward sustainment and capacity development for the electricity sector, which included the award of a long-term operations and maintenance support contract to (1) support Ministry of Electricity efforts to develop, implement, and sustain a functional and effective operations and maintenance organization and plan, and (2) perform operations and maintenance activities at select thermal and gas turbine power facilities. According to DOD, GRD also received $250 million in fiscal year 2006 Economic Support Funds to support sustainment and capacity development programs in the electricity sector.

MAJOR CHALLENGES HINDER EFFORTS TO MEET IRAQ'S OIL AND ELECTRICITY NEEDS

The U.S. government and Iraq face several key challenges in improving Iraq's oil and electricity sectors. First, the U.S. reconstruction program assumed a permissive security environment that never materialized; the ensuing lack of security resulted in project delays and increased costs. Second, corruption, smuggling, and other illicit activities have diverted government revenues potentially available for rebuilding efforts. Third, according to the World Bank, Iraq's lack of an adequate legal and regulatory framework and financial management systems hinder progress and make it difficult to attract foreign investors. Finally, although the oil and electricity sectors are mutually dependent, the Iraqi government lacks integrated planning for these sectors leading to inefficiencies that could hinder future rebuilding efforts.

POOR SECURITY CONDITIONS HAVE SLOWED RECONSTRUCTION AND INCREASED COSTS

The U.S. reconstruction effort was predicated on the assumption that a permissive security environment would exist. However, since June 2003, overall security conditions in Iraq have deteriorated and grown more complex, as evidenced by the increased numbers of attacks and the Sunni-Shi'a sectarian strife that followed the February 2006 bombing of the Golden Mosque in Samarra (see fig. 8). Enemy-initiated attacks against the coalition and its Iraqi partners continued to increase through October 2006 and remain high. The

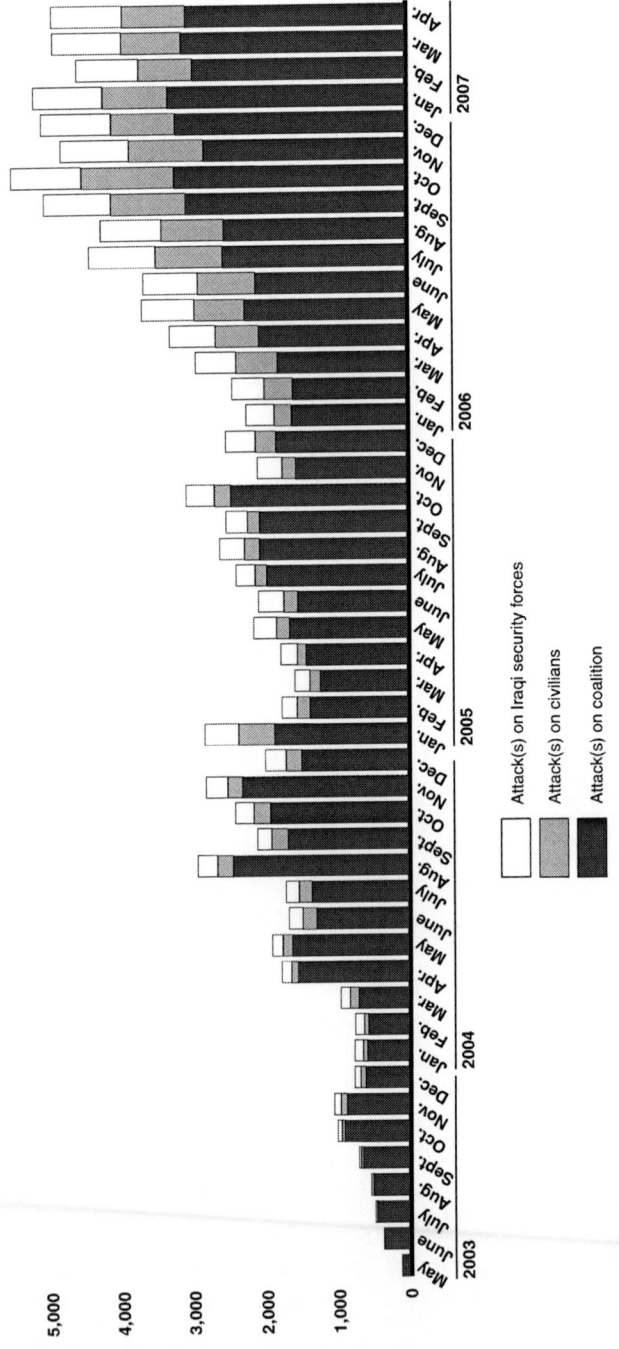

Figure 8. Enemy-Initiated Attacks against the Coalition and Its Iraqi Partners.

Sources: Multi-National Force-Iraq, April 2007.

average total number of attacks per day has risen from 71 per day in January 2006 to a record high of 176 per day in October 2006. For the last 3 months, average attacks per day were 164 in February, 157 in March, and 163 in April 2007.

The deteriorating security environment continues to pose a serious challenge to Iraq's oil and electricity sectors and has, in part, led to project delays and increased costs. Insurgents have destroyed key oil and electricity infrastructure (see fig. 9), threatened workers, compromised the transport of materials, and hindered project completion and repairs by preventing access to work sites. Moreover, looting and vandalism have continued since 2003. U.S. officials reported that major oil pipelines in the north continue to be sabotaged, shutting down oil exports and resulting in lost revenues (see fig. 10). For example, according to GRD, although eight gas oil separation plants in northern Iraq have been refurbished, many are not running due to interdictions on the Iraq-Turkey pipeline and new stabilization plant. GRD noted that if the lines and plant were in operation today, an additional 500,000 barrels per day would be produced in northern Iraq. Major electrical transmission and fuel lines also have been repeatedly sabotaged, cutting power to other parts of the country. According to Ministry of Electricity and U.S. officials, workers are frequently intimidated by anti-coalition forces and have difficulty repairing downed lines. Oil pipeline repair crews also are overwhelmed by the amount of work and are unable to make rapid repairs.

Source: State Department's Iraq Reconstruction Management Office.

Figure 9. Downed Transmission Tower in Iraq.

Source: U.S. Army Corps of Engineers Gulf Region Division

Figure 10. Attack on Oil Pipeline.

Although it is difficult to quantify the costs of poor security conditions in time and money, agency and contractor documents and agency officials report numerous security-related issues that have resulted in delays in the design and execution of projects and reduced scopes of work. Poor security conditions also have increased the cost of providing security services for contractors and sites. For example, when a project is shut down or delayed due to security conditions, the fixed costs of contractor camps and salaries continue to accrue even though contractors in the field are unable to continue their work. GAO previously reported that our analyses of five U.S.-funded electricity sector contracts indicate that as of December 31, 2004, security costs to obtain private security services and security-related equipment ranged from 10 to 36 percent of project costs.[30] In December 2006, GAO reported that DOD estimated that its design-build contractors would incur security costs of about 14 percent on electricity contracts worth about $732 million.[31]

In an effort to stop the sabotage of electricity infrastructure, the Ministry of Electricity contracted with tribal chiefs to protect the transmission lines running through their areas, paying them about $60 to $100 per kilometer, according to State's Iraq Reconstruction Management Office (IRMO). However, in October 2006, IRMO officials reported that this scheme was flawed and did not result in improved infrastructure protection. According to U.S. and UN Development Program officials, some tribes that were paid to protect transmission lines also sold materials from the downed lines and extracted tariffs for access to repair the lines.

The U.S. government has developed a number of other initiatives to better protect the oil and electricity infrastructure and transfer this responsibility to the Iraqi government.[32] Such efforts include fortifying the physical integrity of the infrastructure, improving rapid repair capability, and improving the capabilities of infrastructure protection security forces such as the Oil Protection Force, the Electrical Power Security Service (EPSS), and the Strategic Infrastructure Battalions (SIB). The U.S. government has paired these security forces with coalition partners and has trained and equipped the EPSS[33] and SIBs. However, U.S. officials stated that the EPSS effort was unsuccessful and that the capability and loyalty of some of the SIB units are questionable. According to a U.S. official and a recent Center for Strategic and International Studies report,[34] these security forces have been underpaid, underequipped, and poorly led, and are of questionable quality. Additional information on the nature and status of these efforts and the SIBs is classified.

CORRUPTION, SMUGGLING, AND OTHER ILLICIT ACTIVITIES IMPACT REVENUES AND COST RECOVERY

Corruption in Iraq is reportedly widespread and poses a major challenge to building an effective Iraqi government and effective institutions. A World Bank report notes that corruption undermines the government's ability to make effective use of current reconstruction assistance. A 2006 survey by Transparency International ranked Iraq's government as the second most corrupt in the world.

U.S. and international officials have noted that corruption in Iraq's oil sector is pervasive. In 2006, the World Bank and Ministry of Oil's Inspector General estimated that millions of dollars of government revenue are lost each year to oil smuggling or diversion of refined products. According to State Department officials and reports, about 10 percent to 30 percent of refined fuels are diverted to the black market or are smuggled out of Iraq and sold for a profit. According to State Department reporting, Iraqi government officials may have profited from these activities. The insurgency has been partly funded by corrupt activities within Iraq and from skimming profits from black marketers, according to U.S. Embassy documents. One factor that had stimulated black market activities and fuel smuggling to neighboring countries was Iraq's low domestic fuel prices, which were subsidized by the government. However, under the IMF's Stand-by Arrangement with Iraq, the government has already increased domestic fuel prices five times,[35] significantly reducing the subsidy for many fuel products. The Iraqi government intends to continue the price increases during 2007, as well as encourage private importation of fuels, which was liberalized in 2006. The purpose is to decrease the incentive for black market smuggling and to increase the availability of fuel products.

Illegal connections to existing power lines and nonfunctioning meters negatively affect the Iraqi Ministry of Electricity's ability to manage its electricity system and finance the improvements needed. According to U.S. and international donor officials, electricity meters may be old, intentionally damaged, inaccurate, nonfunctioning, or nonexistant for some businesses and households. As a result of illegal connections and lack of metering, the Ministry of Electricity cannot obtain an accurate measure of the electricity consumed by households and businesses, which contributes to inaccurate billing and low cost recovery.

INADEQUATE LEGAL AND REGULATORY FRAMEWORK AND FINANCIAL MANAGEMENT SYSTEMS HINDER PROGRESS

According to the World Bank, Iraq lacks clear institutional, legal, and regulatory structures and adequate financial management systems. These would ensure more effective resource management, transparency of oil revenues, and the financial accountability of Iraqi ministry and plant managers.

The World Bank reports that additional incentives are needed to stimulate oil production and investment. According to the Bank, these incentives would include a clear legal and regulatory framework, clearly assigned roles for Iraq's ministries, state agencies, and the private sector, and a predictable negotiating environment for contracts. According to U.S.

officials, until a new hydrocarbon (oil and gas) law is passed and implemented, uncertainties exist for the oil sector in each of these areas.

As of February 26, 2007, the Iraqi Cabinet (Council of Ministers) had approved a draft oil and gas law that as of May 1, 2007 awaits consideration and enactment by the Parliament (Council of Representatives). The draft law was expected to clearly assign roles, decentralize the development of oil and gas fields, centralize control of revenues, and grant regions[36] and regional oil companies the right to draw up contracts with foreign companies for exploration and development of new oil fields.[37] However, according to State Department officials, although the draft oil and gas law provides a necessary framework, some vital provisions, which will be in four annexes and companion legislation,[38] have yet to be prepared. Moreover, according to the State Department, the oil and gas law cannot go forward without the companion legislation. According to the Kurdistan Regional Government, the Oil and Energy Committee of the Iraq Council of Ministers will prepare the annexes, which will allocate the management of particular petroleum fields and exploration areas in Iraq to the Kurdistan Regional Government, the Iraq National Oil Company, and the Iraq Ministry of Oil. According to State Department and Kurdistan Regional Government officials, the Committee also needs to come to agreement on model petroleum contracts and guidelines for contractual terms. The Kurdistan Regional Government has stated that, until this is completed, no investment in Iraq can begin.

According to U.S. officials, the oil and gas law is necessary but not sufficient to attract the near-term involvement of international oil companies. The legislation does not clearly detail how revenue resulting from foreign participation will be divided. According to State Department officials, the revenues will be collected in a central account and shared among provinces on the basis of population levels, but the details will be addressed in a separate hydrocarbon financial management law. Some U.S. officials note that the per capita distribution of funds will require a politically sensitive census to be undertaken.

After the Iraqi government passes legislation to enhance the transparency of the legal and regulatory environment, it sometimes has difficulty developing related implementing regulations and guidelines. For example, according to the Department of State, implementing regulations have yet to be issued for Iraq's Fuel Import Liberalization Law, which passed on September 6, 2006.[39] According to the State Department, the law allows private imports of refined fuel products and thereby breaks the monopoly of government imports and sets the stage for the private importation of gasoline and diesel at market prices. According to U.S.

officials, the Ministry of Oil and the Iraqi National Oil Company may have had difficulty recruiting the skilled oil and gas technical experts needed to craft implementing regulations.

Additionally, according to U.S. and donor officials, the Iraqi government lacks adequate financial management systems at all levels that would ensure the financial accountability of Iraqi ministry and plant managers. The World Bank reported that effective oil generation and management include transparency in collecting, monitoring, and reporting revenues. U.S. and international officials report that, given the importance of oil revenues to the Iraqi economy, management of these revenues is crucial.[40] The United States and international donors are working with the Iraqi government to improve financial management practices at Iraq's key ministries, including the oil and electricity ministries.

LACK OF INTEGRATED ENERGY PLANNING CREATES INEFFICIENCIES AND COULD HINDER FUTURE REBUILDING EFFORTS

Although the oil and electricity sectors are mutually dependent, the Iraqi government lacks integrated planning for these sectors, according to U.S. and international donor community officials. Specifically, the Iraqi government lacks an integrated energy plan that clearly identifies future costs and resource needs; rebuilding goals, objectives, and priorities; stakeholder roles and responsibilities, including steps to ensure coordination of ministerial and donor efforts; an assessment of the environmental risks and threats; and performance measures and milestones to monitor and gauge progress. For example, the lack of cooperation and coordination among the Oil and Electricity Ministries, particularly in supplying appropriate fuels to the electricity sector, has resulted in inefficiencies such as increased maintenance costs and frequent interruptions in electricity production, according to IRMO officials. According to U.S. and international donor officials, the Ministries of Oil and Electricity will need to more closely coordinate their efforts to achieve the Iraqi government's ambitious future rebuilding goals.

According to IRMO advisors to the Iraqi government, a reliable supply of electricity is essential to the oil industry and vice versa. The electricity sector is heavily dependent on fuels produced by the oil sector, and oil and gas facilities rely on electricity supplied from the national grid. For example, in March 2007, IRMO reported that a problem within the national electricity grid caused a loss of power to the Bayji oil refinery, Iraq's largest refinery, resulting in a reduced output of refined fuel products. As we previously reported, Iraq's fuel supply does not meet demand and is of inconsistent quality, resulting in inefficient operation of the gas turbine electricity generators that help power the national grid.

Although the need for an integrated plan has been recognized, an integrated energy sector plan has yet to be developed and the current electricity master plan does not state how future coordination will take place and will require actions that have been difficult to achieve in the past. The Iraqi Minister of Electricity recognized the need for integrated planning in his speech at the November 2006 UNDP and Iraqi government electricity conference at which the 2006-2015 Electricity Master Plan was presented to donors. According to the Minister, the energy sector is faced with the challenge of expediting simultaneous implementation of strategic projects, since one sector complements the other. However, the Ministry of Oil's contribution to the Electricity Master Plan was limited to a description of the current fuel situation and the types of fuels the Ministry of Oil would make available for power generation. According to an IRMO advisor to the Ministry of Oil, the Ministry of Electricity merely includes the requirements for supplying electricity to oil facilities within its annual power budget. Moreover, international donors cautioned that the 2006-2015 Electricity Master Plan does not address how future coordination between the two ministries will be ensured or how events affecting one sector will be mitigated to prevent disruptions in the other. Officials of one key donor country also noted that the rigor used to prioritize electricity projects was questionable. According to the World Bank, Iraq needs sound economic criteria to help evaluate whether to rehabilitate or renew electricity projects.

Further, international donors stated that, although the Electricity Master Plan contains milestones, it is ambitious, given the current condition of the infrastructure, sizable funding need, and limited past coordination between the Electricity and Oil Ministries. According to the UNDP, the plan's successful implementation will require actions that have been difficult to achieve in the past such as the coordination of all stakeholders, mobilization and timely release of funds, conversion of heavy fuel oil to natural gas, building of new generating capacity closer to demand, and a reduction in sabotage and looting.

According to IRMO officials, the Ministry of Oil has not developed a plan comparable to the Electricity Master Plan and currently has no plan to detail how it will increase the country's crude oil production capacity to reach or surpass 6 million barrels per day by 2015. According to GRD, Iraq's current oil production capabilities remain unknown and no master plans exist for actions that will enable Iraq to increase its capacity in the future. According to the U.S. government, Iraq is losing about $4.1 billion per year in potential fuel revenues through its inefficient energy production practices, such as the flaring of natural gas, underscoring the need for such a plan.

CONCLUSION

The United States has spent billions of dollars to rebuild Iraq's oil and electricity sectors. However, billions more will be needed to surmount the current security, corruption, management, and legal challenges facing Iraq and attain the U.S. program goals first established under the CPA. The absence of reliable crude oil production and refinery data complicate efforts to measure progress and identify trends in government revenues and possible illicit diversions. Weak budgeting and procurement practices have also slowed Iraq ministry efforts to spend funds already approved for reconstruction projects. Although the United States is exploring some new options for providing rebuilding services on an advance-of-funds basis, it is unclear how this arrangement will contribute to the improvement of ministry procurement and budgeting systems. In addition, the lack of a clear legal and regulatory environment impedes new foreign investment. The passage of Iraqi hydrocarbon legislation could serve as an important impetus for stimulating additional investment if and when security conditions improve to acceptable levels. However, it is difficult to identify the most pressing future funding needs, key rebuilding priorities, and existing vulnerabilities and risks given the absence of an overarching energy sector strategic plan that comprehensively assesses the requirements of the energy sector as a whole. Given the highly interdependent nature of the oil and electricity sectors, such a plan would help identify the most pressing needs for the entire energy sector and help overcome the daunting challenges affecting future development prospects.

RECOMMENDATIONS FOR EXECUTIVE ACTION

We recommend that the Secretary of State, in conjunction with relevant U.S. agencies and in coordination with the donor community, work with the Iraqi government and particularly the Ministries of Oil and Electricity to:

1. Develop an integrated energy strategy for the oil and electricity sectors that identifies and integrates key short-term and long-term goals and priorities for rebuilding, maintaining, and securing the infrastructure; funding needs and sources; stakeholder roles and responsibilities, including steps to ensure coordination of ministerial and donor efforts; environmental risks and threats; and performance measures and milestones to monitor and gauge progress. The strategic plan should include an integrated fuel strategy to ensure the delivery of appropriate refined fuels for more efficient electricity production and a risk-based method for prioritizing future projects.
2. Set milestones and assign resources to expedite efforts to establish an effective metering system for the oil sector that will enable the Ministry of Oil to more effectively manage its network and finance improvements through improved measures of production, consumption, revenues, and costs.
3. Improve the existing legal and regulatory framework, for example, by setting milestones and assigning resources to expedite development of viable and equitable hydrocarbon legislation, regulations, and implementing guidelines that will enable effective management and development of the oil sector and result in

increased revenues to fund future development and essential services.
4. Set milestones and assign resources to expedite efforts to develop adequate ministry budgeting, procurement, and financial management systems.
5. Implement a viable donor mechanism to secure funding for Iraq's future oil and electricity rebuilding needs and for sustaining current energy sector infrastructure improvement initiatives once an integrated energy strategic plan has been developed.

List of Congressional Committees

The Honorable Daniel K. Inouye
Chairman
The Honorable Ted Stevens
Ranking Member
Subcommittee on Defense
Committee on Appropriations
United States Senate

The Honorable Patrick J. Leahy
Chairman
The Honorable Judd Gregg
Ranking Member
Subcommittee on State, Foreign Operations,
 and Related Programs
Committee on Appropriations
United States Senate

The Honorable Carl Levin
Chairman
The Honorable John S. McCain
Ranking Member
Committee on Armed Services
United States Senate

The Honorable Joseph R. Biden, Jr.
Chairman

The Honorable Richard G. Lugar
Ranking Member
Committee on Foreign Relations
United States Senate

The Honorable Joseph I. Lieberman
Chairman
The Honorable Susan M. Collins
Ranking Member
Committee on Homeland Security
    and Governmental Affairs
United States Senate

The Honorable John P. Murtha, Jr.
Chairman
The Honorable C.W. Bill Young
Ranking Member
Subcommittee on Defense
Committee on Appropriations
House of Representatives

Nita M. Lowey
Chair
Frank R. Wolf
Ranking Member
Subcommittee on State, Foreign Operations,
    and Related Programs
Committee on Appropriations
House of Representatives

The Honorable Ike Skelton
Chairman
The Honorable Duncan L. Hunter
Ranking Member
Committee on Armed Services
House of Representatives

The Honorable Tom Lantos
Chairman

The Honorable Ileana Ros-Lehtinen
Ranking Member
Committee on Foreign Affairs
House of Representatives

The Honorable Henry A. Waxman
Chairman
The Honorable Tom Davis
Ranking Member
Committee on Oversight and Government Reform
House of Representatives

The Honorable John F. Tierney
Chairman
The Honorable Christopher Shays
Ranking Member
Subcommittee on National Security
 and Foreign Affairs
Committee on Oversight and Government Reform
House of Representatives

APPENDIX I: OBJECTIVES, SCOPE, AND METHODOLOGY

To examine U.S. activities directed at rebuilding the oil and electricity sectors, we assessed (1) the funding made available to rebuild Iraq's oil and electricity sectors and the factors that may affect Iraq's ability to meet its future funding needs, (2) the U.S. goals for the oil and electricity sectors and progress in achieving these goals, and (3) the key challenges the U.S. government faces in helping Iraq restore its oil and electricity sectors. As part of this work, we made multiple visits to Iraq and Jordan during 2006. We conducted our review from January 2006 through March 2007 in accordance with generally accepted government auditing standards. The information on foreign law in this report does not reflect our independent legal analysis but is based on interviews and secondary sources.

To address the amount of U.S. funds that were appropriated, obligated, and disbursed for the reconstruction of Iraq's oil and electricity sectors, we collected funding information from the Department of Defense (DOD), U.S. Army, Department of State, U.S. Agency for International Development (USAID), and the Coalition Provisional Authority (CPA). Data for U.S.-appropriated funds are as of September 30, 2006. We also reviewed U.S. agency inspector generals' reports, Special Inspector General for Iraq Reconstruction (SIGIR) reports, and other audit agency reports. Although we have not audited the funding data and are not expressing our opinion on them, we discussed the sources and limitations of the data with the appropriate officials and checked them, when possible, with other information sources. We determined that the data were sufficiently reliable for broad comparisons in the aggregate, and we noted limitations at the sector level.

To address the amount of Iraqi funds used by U.S. agencies to support the rebuilding of Iraq's oil and electricity sectors, we relied on CPA documents, past GAO interviews with the CPA, and reviews of independent auditor's reports. To identify sources and uses of Development Fund for Iraq (DFI) funds for the oil and electricity sectors, as agreed with U.S. officials, we relied on funding data obtained from

Development Fund for Iraq Statement of Cash and Receipts and Payments (with Independent Auditor's Report) reports for January 1, 2004, through December 31, 2005, from the International Monitoring Advisory Board. We determined the data were sufficiently reliable to describe the major disbursements from the DFI to the oil and electricity sectors and noted limitations in reporting as identified in the independent auditor's report. For the sources and uses of vested and seized assets, we relied on CPA and DOD funding data. To determine the reliability of these data, we examined CPA schedules and relied on past GAO interviews with CPA officials responsible for the data. We verified results with DOD officials. We determined that the data were sufficiently reliable for broad comparisons in the aggregate, and we noted limitations at the sector level.

To assess Iraqi budget allotments and expenditures for the oil and electricity sectors, as well as potential future budget support, we obtained and assessed Iraqi budget data from the U.S. Department of the Treasury. We corroborated our analysis and findings with information from other U.S. agencies, the Iraqi government, and the International Monetary Fund. While we did not independently verify the precision of the data on Iraq's budget execution, we found that they were sufficiently reliable to show the relative differences in budget execution across Iraq's ministries and budget categories (e.g., capital projects versus salaries). Finally, we obtained oil revenue and pricing data from the Department of State, Iraq Reconstruction Management Office (IRMO) officials and discussed with them the collection and preparation of these statistics. We also corroborated the data with external sources. Data on revenue are based on U.S. agency estimates that use internationally recognized financial sources for pricing calculations, such as Bloomberg and Platts. We found that these data were sufficiently reliable for the purpose of showing trends in Iraq's crude oil export revenue and prices.

To address international assistance for the rebuilding of Iraq's oil and electricity sectors, we reviewed documents and met with U.S., Iraqi, and international donor officials. As agreed with the Department of State, United Nations (UN), and World Bank officials, we obtained data and information on multilateral contributions from the UN Development Programme Iraq

Appendix I: Objectives, Scope, and Methodology

Trust Fund and the World Bank Iraq Trust Fund through the Iraq Reconstruction Fund for Iraq Web site. [1] As agreed with Department of State officials, we obtained data and information on bilateral contributions from their quarterly report to Congress[2] and the Iraqi Ministry of Planning's Development Assistance Database (DAD).[3] We also obtained data and information on bilateral contributions from international donors, including Japan and the European Union. To assess the reliability of the data on pledges, commitments, and deposits made by international donors, we interviewed officials at the Department of State who are responsible for monitoring the data provided by the IRFFI and donor nations. We determined that the data on donor commitments and deposits made to the IRFFI were sufficiently reliable for the purposes of reporting at the sector level. Based on discussions with the Ministry of Planning, UN, World Bank, and Department of State officials, we determined that the DAD was incomplete and would not provide accurate data for reporting purposes. However, according to Department of State officials, this is the best available source of data and the department uses this data in their reports to Congress.

To assess U.S. goals for the electricity and oil sectors, U.S. measures of progress in achieving these goals, and the planning and status of reconstruction projects, we reviewed reports and planning documents obtained from the Department of State's IRMO and Near East Asia bureau; DOD's U.S. Army Corps of Engineers Gulf Region Division (GRD) and Project Contracting Office, including contractors; USAID; CPA; U.S. embassy officials in Iraq; International Monetary Fund; the World Bank; and the UN. We also obtained views from international donors such as Japan and the European Community. To assess U.S. goals, we reviewed planning and strategy documents from the CPA and U.S. government, including the National Strategy for Victory in Iraq; Iraq's *National Strategy for Development; the MNF-I/U.S. Embassy Baghdad Joint Action Campaign Plan*; and the Department of *State's Quarterly Update to Congress, 2207 Report*, among others. To monitor the status of efforts and reconstruction projects, we reviewed daily, weekly, and biweekly reporting from the Department of State, DOD, GRD, and USAID. We compiled daily data from IRMO to provide summary data on crude oil and electricity production averages. We discussed and reviewed documents on goal, metric, data limitations, and project status with GRD, USAID, IRMO, and international donor officials. IRMO collects its crude oil production and export data from estimates provided by Iraq's Ministry of Oil and state-owned companies that have responsibility for various regions and parts of the production process.

We identified the challenges IRMO faces in collecting these data, which we discuss in the report, including limited metering at oil facilities. We also compared crude oil production data from IRMO to data reported by the Department of Energy, Energy Information Administration (EIA). We noted a discrepancy between these two data sets and discussed with EIA officials the reasons for this discrepancy, which we note in the report. We also examined IRMO data on Iraq's national stocks of refined petroleum products, such as gasoline. For electricity production, IRMO receives data from Iraq's Ministry of Electricity based on electricity output from Iraq's electricity national grid. It does not include electricity production from local generators. We also reviewed prior GAO, U.S. agency inspector general, SIGIR, and other audit agency reports. Based on this analysis, we found the data sufficiently reliable for identifying production goals in both sectors and whether actual production is meeting these goals. We report oil and electricity production data as of December 31, 2006, and reconstruction project progress data as of March 3, 2007.

To determine the challenges affecting sector progress, we reviewed and analyzed U.S., Iraqi, donor government, UN, International Monetary Fund (IMF), and World Bank reports and data. Over the course of our field work in Washington, D.C.; Baghdad, Iraq; and Amman, Jordan, we met with officials from the Department of State, USAID, DOD, the Department of the Treasury, and the Department of Energy. Over the course of our two trips to Iraq and Jordan, we met with Iraqi, UN, International Monetary Fund, World Bank, donor country (Japan and European Union), and private-sector officials. We observed several embassy meetings and a donor meeting led by the Iraqi Ministry of Planning and Development. We also interviewed U.S., Iraqi, and UN officials at a November 2006 electricity conference sponsored by the UN Development Program at the Dead Sea, Jordan.

REFERENCES

[1] These figures include the Iraq Relief and Reconstruction Funds provided in the Emergency Wartime Supplemental Appropriations Act, 2003, P.L. 108-11, and in the Emergency Supplemental Appropriations Act for Defense and for the Reconstruction of Iraq and Afghanistan, 2004, P.L. 108-106, which are known as IRRF 1 and IRRF 2, respectively.

[2] In this report, the oil sector, which is under the direction of Iraq's Ministry of Oil, refers broadly to crude oil, associated natural gas production and exports, and petroleum products (such as gasoline) that are refined in Iraq or imported.

[3] GAO, *Rebuilding Iraq: Governance, Security, Reconstruction, and Financing Challenges*, GAO-06-697T (Washington, D.C.: Apr. 25, 2006); *Rebuilding Iraq: More Comprehensive National Strategy Needed to Help Achieve U.S. Goals*, GAO-06-788 (Washington, D.C.: July 11, 2006); *Rebuilding Iraq: More Comprehensive National Strategy Needed to Help Achieve U.S. Goals and Overcome Challenges*, GAO-06-953T (Washington, D.C.: July 11, 2006); and *Securing, Stabilizing, and Rebuilding Iraq: Key Issues for Congressional Oversight*, GAO-07-308SP (Washington, D.C.: Jan. 9, 2007).

[4] The first year of data that the U.S. Department of Energy's Energy Information Administration (EIA) reports for Iraq is 1970.

[5] UN/World Bank, *Joint Iraq Needs Assessment* (Baghdad, Iraq: Oct. 2003).

[6] GAO, *Rebuilding Iraq: More Comprehensive National Strategy Needed to Help Achieve U.S. Goals*, GAO-06-788 (Washington, D.C.: July 2006).

[7] Funding for CERP was made available in the following legislation: Emergency Supplemental Appropriations Act for Defense and for the Reconstruction of Iraq and Afghanistan, 2004, P.L. 108-106; Department of Defense Appropriations Act, 2005, P.L. 108-287; Emergency Supplemental Appropriations Act for Defense, the Global War on Terror, and Tsunami Relief, 2005, P.L. 109-13; Department of Defense, Emergency Supplemental Appropriations to Address Hurricanes in the Gulf of Mexico, and Pandemic Influenza Act, 2006, P.L. 109-148; and Emergency Supplemental Appropriations Act for Defense, the Global War on Terror, and Hurricane Recovery, 2006, P.L. 109-234.

[8] P.L. 109-289.

[9] "Vested assets" refers to former Iraqi regime assets held in U.S. financial institutions that the President confiscated in March 2003 and vested in the U.S. Treasury. The United States froze these assets shortly before the first Gulf War. The U.S.A. PATRIOT Act of 2001 amended the International Emergency Economic Powers Act to empower the President to confiscate, or take ownership of, certain property of designated entities, including these assets, and vest ownership in an agency or individual. The President has the authority to use the assets in the interests of the United States. In this case, the President vested the assets in March 2003 and made these funds available for the reconstruction of Iraq in May 2003. "Seized assets" refers to former regime assets seized within Iraq.

[10] On May 22, 2003, the UN adopted Security Council Resolution 1483, which noted the establishment of the DFI to be used for the economic reconstruction and repair of Iraq's infrastructure, among other purposes. Under the resolution, 95 percent of oil proceeds are to be deposited into the DFI.

[11] According to an independent audit of the DFI, U.S. agencies did not maintain a complete and accurate database of outstanding contractual commitments for contracts signed by the former CPA. See Ernst & Young, *Management Letter for Development fund for Iraq for the Period from July 1, 2005, to December 31, 2005* (July 10, 2006).

[12] The CPA turned DFI stewardship over to the new Iraqi government in June 2004. According to State Department estimates, about $18 billion in oil revenue had been deposited into the DFI since the transition from the CPA to the interim Iraqi government, as of May 31, 2005.

[13] While the limited spending by various Iraqi ministries in certain areas, such as capital projects, is a known issue and receiving attention from

U.S. advisors, we cannot verify the precision of these numbers. For the purpose of this report, we only use these data to identify limited spending as a potential challenge for Iraq should it rely on its ministries' own budgets to fund future reconstruction projects. IRMO also stated that the Ministry of Electricity obligation rate was 78 percent. However, IRMO did not provide supporting documentation nor did it provide obligation data for the Ministry of Oil.

[14] According to the State Department, the Contracting Committee requirement for about a dozen signatures to approve electricity and oil contracts exceeding $10 million further slows a bureaucratic process.
[15] World Bank, *Operation Procurement Review* (June 2005).
[16] The established mechanism for channeling multilateral assistance to Iraq is the International Reconstruction Fund Facility for Iraq (IRFFI), which is composed of two trust funds, one run by the UN Development Group and the other by the World Bank Group.
[17] The Iraqi government collects self-reported donor funding data through its Donor Assistance Database (DAD). Iraqi and U.S. government and international donor country officials agree that DAD is incomplete. According to the Department of State and SIGIR officials, the Ministry of Planning's DAD, supported by the UN Development Programme (UNDP), continues to improve as an assistance management tool to track donor assistance.
[18] Department of State, *Quarterly Update to Congress; 2207 Report* (Washington, D.C.: Jan. 2007).
[19] GAO, *Securing, Stabilizing, and Rebuilding Iraq: Key Issues for Congressional Oversight*, GAO-07-308SP (Washington, D.C.: Jan. 2007).
[20] U.S. goals differ from the government of Iraq and IMF targets. According to the State Department, as of January 2007, Iraq's production goal is 2.1 mbpd and its export goal is 1.7 mbpd.
[21] In Iraq, most natural gas and liquefied petroleum gas production is associated with production of crude oil. As crude oil is pumped from the ground and processed, natural gas and liquefied petroleum gas are also produced. Refined petroleum products are produced in refineries in Iraq from crude oil sent through pipelines from the oil fields or are imported from other countries since Iraq's refinery capacity does not currently meet national needs.
[22] EIA uses its own analysis and a variety of sources, including Dow Jones, the *Middle East Economic Survey*, the *Petroleum Intelligence*

Weekly, the International Energy Agency (IEA), the *Monthly Oil Market Report* from OPEC, the *Oil & Gas Journal*, Platts, and Reuters.

[23] Re-injecting crude oil, and more commonly heavy fuel oil (residual oil), back into wells is one way to maintain pressure in the wells for extraction. However, this also can damage the wells and reduce the value of the remaining reserves.

[24] The State Department set a goal to attain 6,000 mw peak generation capacity by July 2006. The State Department also set a goal of 6,137 mw average peak daily supply for the summer of 2006, which was also not met.

[25] A megawatt hour is a measure of energy production or consumption equal to megawatts produced or consumed for 1 hour.

[26] The State Department set an 110,000 mwh average daily generation goal for the summer of 2005. For the summer of 2006, the State Department set a goal of 127,000 mwh. However, the State Department continues to use the 110,000 mwh goal to track weekly progress. According to the State Department, generally goals have been announced and changed without a formal process.

[27] Reporting average daily load (megawatt hours) is a better measure of how much power is produced for the national grid because it measures generation over a period of time, rather than the peak (megawatt) for the day.

[28] See Emergency Supplemental Appropriations Act for the Defense and Reconstruction of Iraq and Afghanistan, 2004, P.L. 108-106.

[29] "Workover" is a term used to describe operations on a completed production well to clean, repair, and maintain the well for the purposes of increasing or restoring production.

[30] GAO, *Rebuilding Iraq: Status of Funding and Reconstruction Efforts*, GAO-05-876 (Washington, D.C.: July 28, 2005).

[31] GAO, *Rebuilding Iraq: Status of DOD's Reconstruction Program*, GAO-07-30R (Washington, D.C.: Dec. 15, 2006).

[32] The Special Inspector General for Iraq Reconstruction (SIGIR), *Unclassified Summary of SIGIR's Review of Efforts to Increase Iraq's Capability to Protect Its Energy Infrastructure*, SIGIR-06-038 (Washington, D.C.: Sept. 27, 2006).

[33] In March 2004, the United States awarded a $19 million contract to train and equip Iraq's Electrical Power Security Service to protect electrical infrastructure, including power plants, transmission lines, and Ministry of Electricity officials. Although the program was designed to

train 6,000 guards over a 2-year period, fewer than 340 guards had been trained when the contact was terminated early.

[34] Anthony Cordesman, Center for Strategic and International Studies, *Iraqi Force Development and the Challenge of Civil War: the Critical Problems and Failures the U.S. Must Address If Iraqi Forces Are to Eventually Do the Job* (Washington, D.C.: Nov. 30, 2006).

[35] According to State Department officials, the most recent fuel prices were increased on January 5 and March 20, 2007.

[36] According to an English translation of the draft law, "region" is defined as the Region of Kurdistan or any other region formed after issuance of the law in the Republic of Iraq, according to the constitution of Iraq.

[37] According to State Department officials, the draft law arranges for the re-establishment of the Iraqi National Oil Company, which would have control over currently producing oil fields. The development of currently non-producing fields would be subject to an open bidding process. Contract negotiations for these fields would be the responsibility of the regional authorities and would be based on production and development plans, bidding rules, and model contracts set by a new Federal Oil and Gas Council. The Council would also be responsible for final approval of contracts. The Council would be the final arbitrator and would approve all Ministry of Oil strategies and policies.

[38] According to State Department officials, companion legislation will include laws that address the reconstitution of the National oil company, reorganization of the Ministry of Oil, and division of oil revenues.

[39] U.S. Department of State, *Section 1227 Report on Iraq* (Washington, D.C.: Jan. 5, 2007).

[40] The World Bank and UN report that the economic performance of oil exporters is often inferior to that of resource-poor countries. This is often explained by the influence of oil wealth on governance and by the harmful real exchange rate effects on the non-oil sector and uncertainty created by volatile oil prices.

APPENDIX I

[1] The UN Development Fund Iraq Trust Fund and World Bank Trust Fund make up the International Reconstruction Fund Facility for Iraq

(IRFFI). Data and information for UNDP and World Bank funds and programs are provided separately on the IRFFI Web site (see http://www.irffi.org).
[2] The Department of State, Quarterly Update to Congress; 2207 Report (Washington, D.C.: Oct. 2006).
[3] The Iraqi Ministry of Planning Donor Assistance Database is available at http://www.mop-iraq.org/dad.

INDEX

A

access, 37, 39
accountability, 40, 42
accounting, 20
achievement, 17
Afghanistan, 14, 55, 56, 58
agriculture, 21
Army Corps of Engineers, 4, 8, 10, 12, 31, 38, 53
Asia, 53
assessment, 42
assets, 16, 52, 56
assumptions, viii, 11
attacks, 7, 35, 37
attention, 56
auditing, 2, 51
authority, 1, 11, 15, 16, 56
availability, 7, 25, 40

B

Baghdad, 2, 29, 53, 54, 55
basic services, 1
benzene, 25
black market, 5, 39

C

capacity building, 12, 33
cell, 6
Civil War, 59
Coalition Provisional Authority (CPA), x, 11, 12, 13, 16, 18, 24, 25, 28, 29, 45, 51, 52, 53, 56
Commander's Emergency Response Program (CERP), x, 14, 15, 16, 56
Committee on Appropriations, 48, 49
Committee on Armed Services, 48, 49
Committee on Homeland Security, 49
commodities, 19
community, vii, 3, 5, 11, 12, 13, 21, 42, 47
conflict, 1, 11
confusion, 18
Congress, 53, 57, 60
construction, 15, 32
consumers, 17, 25
consumption, 5, 47, 58
contracts, vii, 3, 12, 13, 16, 38, 40, 41, 56, 57, 59
control, 11, 14, 33, 41, 59
conversion, 43
corruption, ix, 4, 18, 26, 35, 39, 45
costs, 4, 11, 35, 37, 38, 42, 47
Council of Ministers, 41

crude oil, ix, 1, 4, 7, 9, 18, 20, 23, 24, 25, 26, 28, 31, 32, 43, 45, 52, 53, 55, 57, 58
cycles, 33

D

data set, 26, 54
database, 56
debt, vii, 3, 21
degradation, 11
delivery, 47
demand, ix, 4, 17, 23, 24, 25, 29, 31, 33, 43
Department of Defense (DOD), x, 2, 12, 14, 15, 16, 17, 18, 19, 31, 33, 34, 38, 51, 52, 53, 54, 56, 58
Department of Energy, 2, 9, 24, 54, 55
Department of State, 2, 5, 12, 30, 41, 51, 52, 53, 54, 57, 59, 60
deposits, 53
Development Fund for Iraq (DFI), x, 16, 28, 52, 56
disbursement, 21
disseminate, 26
distribution, ix, 1, 5, 11, 14, 17, 41
division, 59
domestic demand, 26
Donor Assistance Database (DAD), x, 53, 57, 60
donors, vii, viii, 3, 20, 42, 43, 53
draft, 5, 41, 59

E

East Asia, 53
economic activity, 4
economic assistance, 12
economic development, 4, 21
economic performance, 59
economic resources, 33
electrical power, 19, 33

Electrical Power Security Service (EPSS), x, 39
electricity, vii, viii, ix, 1, 2, 3, 4, 5, 6, 11, 12, 13, 14, 15, 16, 17, 18, 20, 21, 23, 29, 31, 33, 34, 35, 37, 38, 39, 40, 42, 43, 45, 47, 48, 51, 52, 53, 54, 57
electricity system, 40
Emergency Supplemental Appropriations Act, 55, 56, 58
energy, viii, 5, 17, 25, 26, 42, 43, 44, 45, 47, 48, 58
Energy Information Administration (EIA), x, 2, 9, 25, 26, 27, 54, 55, 57
engagement, 21
environment, viii, ix, 4, 11, 28, 31, 35, 37, 40, 41, 45
equipment, 7, 28, 31, 32, 33, 38
estimating, 28
European Community, 53
European Union, 2, 53, 54
exchange rate, 59
execution, 18, 38, 52
expenditures, 2, 19, 52
expertise, 12, 33
exports, 1, 2, 4, 20, 23, 24, 25, 28, 37, 55
extraction, 58

F

failure, 33
fear, 18
feet, x, 25
finance, 11, 40, 47
financial institutions, 56
financing, 7, 21
fixed costs, 38
fluctuations, 20
foreign investment, ix, 5, 45
forgiveness, 21
fuel, ix, 4, 19, 23, 24, 25, 31, 33, 37, 40, 41, 43, 44, 47, 58, 59
funding, vii, viii, 1, 3, 5, 13, 14, 15, 19, 20, 31, 32, 43, 45, 47, 48, 51, 52, 57

Index

funds, vii, viii, 1, 3, 4, 13, 14, 15, 16, 17, 18, 19, 20, 31, 33, 34, 41, 43, 45, 51, 52, 56, 57, 60

G

GAO, viii, 1, 2, 8, 9, 10, 16, 19, 25, 29, 38, 52, 54, 55, 57, 58
gas turbine, 4, 33, 34, 43
gasoline, 5, 25, 41, 54, 55
gauge, 42, 47
generation, ix, 4, 11, 14, 17, 19, 29, 31, 32, 33, 42, 43, 58
goals, viii, ix, 1, 2, 4, 5, 6, 11, 13, 17, 18, 23, 24, 25, 29, 31, 32, 33, 42, 45, 47, 51, 53, 57, 58
governance, 11, 12, 59
government, vii, viii, ix, 1, 2, 3, 4, 6, 11, 12, 13, 16, 17, 18, 19, 20, 21, 23, 26, 28, 33, 34, 35, 39, 41, 42, 43, 44, 45, 47, 51, 52, 53, 54, 56, 57
government revenues, viii, 4, 20, 35, 45
grants, vii, 3, 20
gross domestic product, viii, 1
guidelines, 5, 41, 47
Gulf of Mexico, 56
Gulf Regional Division (GRD), x, 4, 31, 32, 34, 37, 44, 53
Gulf War, 7, 11, 56

H

Homeland Security, 49
House, 49, 50
households, 40
Hussein, Saddam, 18, 21, 26
hydrocarbons, 9

I

implementation, 28, 43
imports, 26, 41
incentives, 40

industry, 17, 28, 42
infrastructure, ix, 1, 4, 7, 11, 15, 17, 23, 31, 33, 37, 38, 39, 43, 47, 48, 56, 58
institutions, 39, 56
integrity, 39
international, vii, viii, ix, 1, 3, 5, 11, 13, 21, 28, 39, 40, 41, 42, 43, 52, 53, 57
International Development Assistance (IDA), x
International Monetary Fund (IMF), x, 2, 20, 26, 40, 52, 53, 54, 57
International Reconstruction Fund Facility for Iraq (IRFFI), x, 53, 57, 60
investment, ix, 5, 17, 20, 40, 41, 45
investors, 20, 35
Iran, 7
Iraq, vii, viii, ix, x, 1, 2, 3, 4, 5, 7, 9, 10, 11, 12, 13, 14, 15, 16, 17, 18, 19, 20, 21, 23, 24, 25, 26, 27, 28, 29, 30, 31, 32, 33, 35, 36, 37, 39, 40, 41, 42, 43, 44, 45, 48, 51, 52, 53, 54, 55, 56, 57, 58, 59
Iraq Reconstruction Management Office (IRMO), x, 2, 4, 12, 19, 26, 27, 28, 31, 32, 39, 42, 43, 52, 53, 57
Iraq Relief and Reconstruction Fund (IRRF), x, 15, 31, 33, 34, 55
Iraq War, 7
Islam, 20

J

Japan, 2, 21, 53, 54
Jordan, 2, 51, 54

K

kerosene, 25
Kuwait, 7

L

laws, 18, 59

legal, iv
legislation, ix, 5, 41, 45, 47, 56, 59
liquefied petroleum gas (LPG), x, 16, 25, 57
liquids, 26
loans, vii, 3, 20, 21
loyalty, 39

M

management, vii, ix, 3, 5, 12, 18, 33, 35, 40, 41, 42, 45, 47, 48, 57
management practices, 42
market, 5, 40, 41
market prices, 41
measurement, ix, 4
measures, 6, 42, 47, 53, 58
megawatt (mw), x, 4, 29, 31, 32, 33, 58
megawatt hours (mwh), x, 29, 58
Mexico, 56
Middle East, 57
military, 12, 14, 15
million barrels per day (mbpd), x, 4, 7, 11, 18, 24, 25, 31, 44, 57
million standard cubic feet per day (mscfd), x, 25
Ministry of Oil, vii, ix, 3, 18, 19, 24, 27, 28, 39, 41, 42, 43, 47, 53, 55, 57, 59
money, 38
monopoly, 41

N

natural gas, 4, 5, 25, 26, 33, 43, 44, 55, 57
neglect, 1, 11
negotiating, 40
network, 7, 11, 28, 47

O

obligation, 57

oil(s), vii, viii, ix, 1, 2, 3, 4, 5, 6, 7, 9, 11, 13, 14, 15, 16, 17, 18, 20, 21, 23, 24, 25, 26, 28, 31, 32, 33, 35, 37, 39, 40, 41, 42, 43, 44, 45, 47, 48, 51, 52, 53, 55, 56, 57, 58, 59
oil production, ix, 1, 2, 4, 7, 9, 11, 18, 24, 25, 26, 28, 31, 32, 40, 44, 45, 53
oil revenues, ix, 5, 11, 13, 16, 17, 20, 40, 42, 59
OPEC, 58
Operation Iraqi Freedom, 7
organization, 12, 34
ownership, 56

P

Pandemic Influenza Act, 56
Parliament, 41
peak demand, 29
perceptions, 1
performance, 6, 11, 32, 42, 47, 59
petroleum, 57
petroleum products, 16, 25, 54, 55, 57
planning, ix, 4, 12, 35, 42, 43, 53
plants, 7, 17, 33, 37, 58
pollution, 33
poor, ix, 4, 23, 31, 32, 33, 38, 59
population, 41
ports, 7
power, 4, 11, 17, 19, 29, 33, 34, 37, 40, 43, 58
power generation, 17, 19, 33, 43
power plants, 58
pressure, 32, 58
prices, 17, 20, 26, 40, 41, 52, 59
private investment, 17
private sector, 2, 40
production, ix, 1, 2, 4, 7, 9, 11, 14, 17, 23, 24, 25, 26, 28, 31, 32, 40, 42, 43, 45, 47, 53, 55, 57, 58, 59
production targets, 31
productivity, 1
profit(s), 5, 40

program, viii, ix, 4, 7, 12, 15, 16, 17, 20, 23, 24, 29, 31, 32, 33, 35, 45, 58
Project and Contracting Office (PCO), x, 12
promote, 21
public investment, 17, 20
pumps, 32

R

range, 20
real time, 26
reconciliation, 5
reconstruction, viii, 1, 4, 6, 11, 12, 13, 14, 15, 16, 17, 18, 20, 21, 23, 24, 29, 35, 39, 45, 51, 53, 56, 57
recovery, ix, 4, 40
recruiting, 42
redevelopment, 21
reduction, 21, 43
refinery capacity, 26, 57
reforms, 21
regional, 41, 59
regulations, 5, 18, 41, 47
regulatory framework, 5, 11, 35, 40, 47
rehabilitation, 14
reliability, 52, 53
repair, 28, 37, 39, 56, 58
reserves, 1, 7, 58
resolution, 56
resource management, 40
resources, ix, 4, 5, 12, 28, 33, 47, 48
responsibility, iv
revenue, ix, 1, 4, 20, 26, 39, 41, 52, 56
risk, ix, 4, 47
rust, 59

S

sabotage, 26, 38, 43
sanctions, 1, 7, 26, 28
Secretary of State, viii, 5, 47

security, viii, ix, 4, 11, 19, 21, 23, 28, 32, 35, 37, 38, 39, 45
Security Council, 56
Senate, 48, 49
separation, 7, 37
series, 26
Shell, 28
sites, 37, 38
skimming, 40
smuggling, ix, 4, 35, 39
Special Inspector General for Iraq (SIGIR), x, 32, 51, 54, 57, 58
speech, 43
stabilization, 6, 12, 37
stakeholder(s), 6, 42, 43, 47
standards, 2, 51
State Department, ix, 4, 5, 10, 12, 15, 19, 21, 24, 25, 27, 29, 37, 39, 41, 56, 57, 58, 59
state-owned enterprises, 28
statistics, 26, 28, 52
stock, 4, 24, 25
storage, 26
Strategic Infrastructure Battalions (SIB), x, 39
strategic planning, 12
strategies, 59
subsidy, 40
summer, 20, 58
supply, ix, 1, 4, 17, 23, 24, 25, 29, 31, 33, 42, 58
surplus, 16
systems, 1, 5, 35, 40, 42, 45, 48

T

targets, 31, 57
tariffs, 39
technical assistance, 21
technology, 7, 33
terminals, 7
theft, 5, 26
threats, 42, 47
time, 7, 25, 26, 38, 58

timing, 26
tons per day (tpd), x, 25, 31
training, 33
transition, 56
translation, 59
transmission, 1, 11, 14, 17, 37, 38, 58
transparency, 19, 40, 41, 42
transport, 37
trend, 26
tribes, 39
trust fund, 57
Turkey, 37
turnover, 18

U

U.S. Agency for International Development, 2, 51
U.S. Department of the Treasury, 52
U.S. Treasury, 19, 56
UN, x, 2, 7, 11, 16, 17, 21, 26, 28, 39, 52, 53, 54, 55, 56, 57, 59
uncertainty, 6, 17, 20, 59
United Kingdom, 11
United Nations, x, 2, 52
United Nations Development Programme (UNDP), x, 2, 43, 57, 60
United States, vii, viii, x, 3, 7, 11, 12, 13, 14, 19, 23, 31, 32, 42, 45, 48, 49, 56, 58
United States Agency for International Development (USAID), x, 2, 12, 15, 31, 32, 51, 53, 54
users, 11

V

values, 9
vandalism, 1, 37
variable, 20
volatility, 13, 17, 20

W

war, 7, 11
War on Terror, 56
Washington, 2, 54, 55, 57, 58, 59, 60
wealth, 59
wells, 7, 32, 58
workers, ix, 4, 33, 37
World Bank, ix, 2, 5, 11, 17, 19, 20, 35, 39, 40, 42, 43, 52, 53, 54, 55, 57, 59